# 7th International Workshop & Summer School on Plasma Physics (IWSSPP'16)



Journal of Physics: Conference Series Volume 982

Kiten, Bulgaria
26 June – 2 July 2016

ISBN: 978-1-5108-5956-2
ISSN: 1742-6588

**Printed from e-media with permission by:**

Curran Associates, Inc.
57 Morehouse Lane
Red Hook, NY 12571



Printed by Curran Associates, Inc. (2018)

For permission requests, please contact the Institute of Physics
at the address below.

Institute of Physics
Dirac House, Temple Back
Bristol BS1 6BE UK

Phone: 44 1 17 929 7481
Fax: 44 1 17 920 0979

techtracking@iop.org

**Additional copies of this publication are available from:**

Curran Associates, Inc.
57 Morehouse Lane
Red Hook, NY 12571 USA
Phone: 845-758-0400
Fax: 845-758-2633
Email: curran@proceedings.com
Web: www.proceedings.com

# Table of contents

## Volume 982

**7th International Workshop and Summer School on Plasma Physics (IWSSPP'16)**

**26 June to 2 July 2016, Kiten, Bulgaria**

**Accepted papers received: 21 February 2018**
**Published online: 8 March 2018**

IOP Conf. Series: Journal of Physics: Conf. Series **982** (2018) 011001 doi:10.1088/1742-6596/982/1/011001

Editorial

# Seventh International Workshop & Summer School on Plasma Physics 2016

Evgenia Benova 2018 J. Phys.: Conf. Ser.

Seventh International Workshop & Summer School on Plasma Physics (IWSSPP'16) was organized by St. Kliment Ohridsky University of Sofia, with co-organizer PLASMER Foundation.

It was held in Kiten, Bulgaria, at the Black Sea Coast, from June 26 to July 2, 2016. The scientific programme covers the topics Fusion Plasma and Materials; Plasma Modeling and Fundamentals; Plasma Sources, Diagnostics and Technology.

The Workshop Plasma Nanoscience was co-organized together with the Institute of Plasmas and Nuclear Fusion, Lisbon, Portugal.

A special Workshop on Remote GOLEM operation was organized by the Institute of Plasma Physics, Prague, Czech Republic for the students and interested participants to work remotely with the Czech TOKAMAK GOLEM.

For the students and interested participants was co-organized one more Workshop “Hunting” for a leak in the vacuum system of a VUV spectrometer, again together with the Institute of Plasmas and Nuclear Fusion, Lisbon, Portugal.

As with the previous issues of this scientific meeting, its aim was to stimulate the creation and support of a new generation of young scientists for further development of plasma physics fundamentals and applications, as well as to ensure an interdisciplinary exchange of views and initiate possible collaborations by bringing together scientists from various branches of plasma physics.

This volume of Journal of Physics: Conference Series includes 14 papers (invited lectures, contributed talks and posters) devoted to various branches of plasma physics, among them fusion plasma and materials, dc and microwave discharge modelling, transport phenomena in gas discharge plasmas, plasma diagnostics, cross sections and rate constants of elementary processes, material processing, plasma-chemistry and technology. Some of them have been presented by internationally known and recognized specialists in their fields; others are MSc or PhD students' first steps in science. In both cases, we believe they will raise readers' interest.

We would like to thank the members of both the International Advisory Committee and the Local Organizing Committee, the participants who sent their manuscripts and passed through the (sometimes heavy and troublesome) refereeing and editing procedure and our referees for their patience and considerable effort to improve the manuscripts. We would like to express our gratitude to the invited lecturers who were willing to pay the participation fee. In this way, in addition to the intellectual support they provided by means of their excellent lectures, they also supported the school financially.

We are obliged to FuseNet Association (http://www.fusenet.eu/) for supporting European master and PhD students' participation in this event and to Fund for Scientific Research at the University of Sofia, Grant No 103/2016.

E. Benova and M. Atanasova

# Peer review statement

All papers published in this volume of *Journal of Physics: Conference Series* have been peer reviewed through processes administered by the proceedings Editors. Reviews were conducted by expert referees to the professional and scientific standards expected of a proceedings journal published by IOP Publishing.

IOP Conf. Series: Journal of Physics: Conf. Series **982** (2018) 012001 doi:10.1088/1742-6596/982/1/012001

# Effect of the resonant magnetic perturbation on the plasma parameters in COMPASS tokamak's divertor region

**M Dimitrova[1,2,5], P Cahyna[1], M Peterka[1,3], E Hasan[2,4], Tsv K Popov[4], P Ivanova[2], E Vasileva[2], R Panek[1], J Cavalier[1], J Seidl[1], T Markovic[1,3], J Havlicek[1], R Dejarnac[1], V Weinzettl[1], P Hacek[1,3], M Tomes[1,3], the COMPASS team and the EUROfusion MST1 team***

[1] Institute of Plasma Physics, The Czech Academy of Sciences, Za Slovankou 3, 182 00 Prague 8, Czech Republic
[2] Emil Djakov Institute of Electronics, Bulgarian Academy of Sciences, 72, Tsarigradsko Chaussee, 1784 Sofia, Bulgaria
[3] Faculty of Mathematics and Physics, Charles University, Ke Karlovu 3, 2116 Prague 2, Czech Republic
[4] Faculty of Physics, St. Kliment Ohridski University of Sofia, 5, James Bourchier Blvd., 1164 Sofia, Bulgaria
* see http://www.euro-fusionscipub.org/mst1

E-mail: dimitrova@ipp.cas.cz

**Abstract**. The resonant magnetic perturbation (RMP) has proven to be a useful way to suppress edge-localized modes that under certain conditions can damage the device by the large power fluxes carried from the bulk plasma to the wall. The effect of RMP on the L-mode plasma parameters in the divertor region of the COMPASS tokamak was studied using the array of 39 Langmuir probes embedded into the divertor target. The current-voltage (*IV*) probe characteristics were processed by the first-derivative probe technique to obtain the plasma potential and the electron energy distribution function (EEDF) which was approximated by a bi-Maxwellian EEDF with a low-energy (4-6 eV) fraction and a high-energy (11-35 eV) one, the both factions having similar electron density. Clear splitting was observed during the RMP pulse in the low-field-side scrape-off-layer profiles of the floating potential $U_{fl}$ and the ion saturation current density $J_{sat}$; these two quantities were obtained both by direct continuous measurement and by evaluation of the *IV* characteristics of probes with swept bias. The negative peaks of $U_{fl}$ induced by RMP spatially overlaps with the local minima of $J_{sat}$ (and $n_e$) rather than with its local maxima which is partly caused by the spatial variation of the plasma potential and partly by the changed shape of the EEDF. The effective temperature of the whole EEDF is not correlated with the negative peaks of $U_{fl}$, and the profile of the parallel power flux density shows secondary maxima due to RMP which mimic those of $J_{sat}$.

## 1. Introduction

During tokamak operation in high energy confinement mode (H-mode), the plasma experiences periodic relaxations of the edge transport barrier known as edge-localized modes (ELMs). The

[5] To whom any correspondence should be addressed.

IOP Conf. Series: Journal of Physics: Conf. Series **982** (2018) 012001 doi:10.1088/1742-6596/982/1/012001

resonant magnetic perturbation (RMP) has proven to be a useful way to suppress ELMs that under certain conditions can damage the device by the large power fluxes carried from the bulk plasma to the wall [1]. Successful mitigation or suppression of the type I ELMs by RMP has already been demonstrated on a number of devices [2-4]. Variations of plasma parameters in the scrape-of layer (SOL) are sensitive indicators of the effects of RMP and its screening by the plasma response [5, 6]. The COMPASS tokamak [7] is equipped with a system of RMP coils with a variety of possible configurations [8], two of which (both with top-bottom symmetry) have recently been utilized for measurements of the magnetic response of ohmic L-mode plasmas to the applied RMP field [9]. Here we report on the effect of this RMP on plasma parameters in the divertor region in these experiments.

The divertor probe system in COMPASS consists of 39 single graphite Langmuir probes embedded in the divertor target and provides profiles of the data measured with a typical spatial resolution in the poloidal direction down to 5 mm. During the experiments, the probes were operated in two different regimes: in a continuous measurement of the floating potential $U_{fl}$, or in a measurement with a swept probe bias. The current-voltage (*IV*) characteristics obtained during the swept regime were processed using the first-derivative probe technique (FDPT) [10-12], which allows evaluation of the plasma potential and the real electron energy distribution function (EEDF).

## 2. Results and discussion

*2.1 Continuous measurement of the $U_{fl}$ by the divertor probes.*

In this regime, the signal from the divertor probes was recorded directly by the COMPASS tokamak DAQ with no additional electronics nor a power supply. As an example, figure 1a presents the temporal profile of the floating potential on the divertor probes, L-mode discharge #9684 (plasma current $I_{pl}$ = 210 kA, toroidal magnetic field $B_T$ = 1.15 T, edge safety factor $q_{95}$ = 3.8 and line-average density $n_e^{avr} = 5\times10^{19}$ m$^{-3}$); the yellow line is the current in the RMP coils (with maximum at +3.5 kA), the black solid lines represent the position of the strike points determined from the magnetic equilibrium reconstruction code EFIT++.

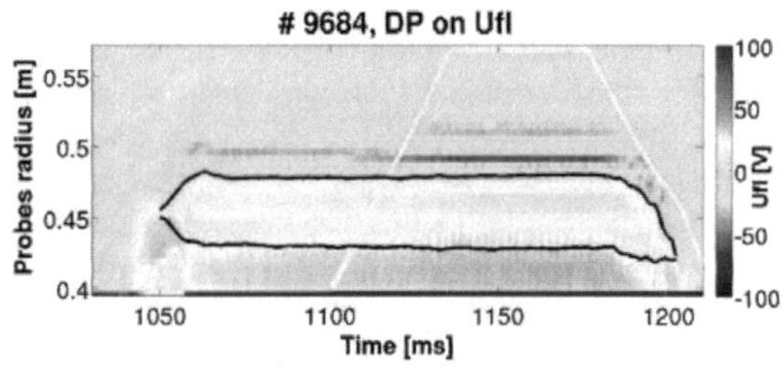


**Figure 1 a.** Temporal evolution of the floating potential on the divertor probes.

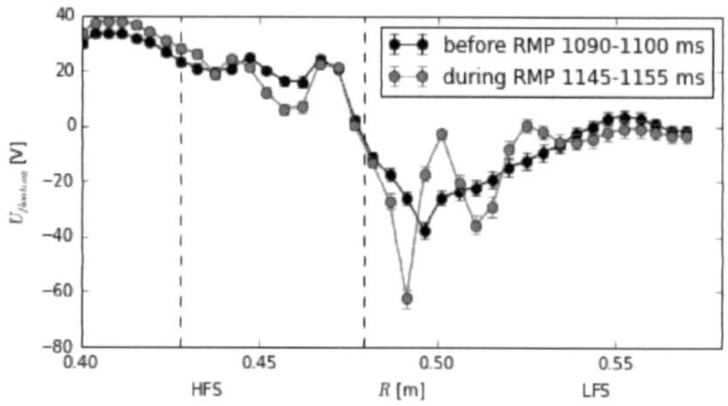


**Figure 1 b.** Poloidal profile of the floating potential before and during RMP in shot #9684.

Figure 1b shows the poloidal profiles of $U_{fl}$ before RMP (black curve) and during RMP at 1150 ms (red curve), the dashed lines indicate the positions of the strike points at 1150 ms. One can see the asymmetry of the floating potential profile – positive values on the high-field side (HFS) and negative on the low-field side (LFS). In contrast to the regime without RMP, during RMP clear splitting of the strike points was observed in the $U_{fl}$ signal as a spatial oscillation in the LFS SOL. This is in qualitative agreement with the striped structure of the visible light emission at the divertor as observed by the fast visible light camera (figure 2) as well as with the lobes intersecting the divertor (figure 3) as predicted by a vacuum model of the last closed flux surface invariant manifolds [13].

The measurements during discharge #9684 were performed at a fixed position of the strike points (black lines, figure 1a). To increase the spatial resolution of the measurements, the X-point was swept, moving the plasma from the LFS towards the HFS. The temporal evolution of the $U_{fl}$ profile during the experiment with swept X-point is shown in figure 4, with the same meaning of yellow and black solid lines as in figure 1a. The profile changes immediately both at the start and the end of the RMP pulse.

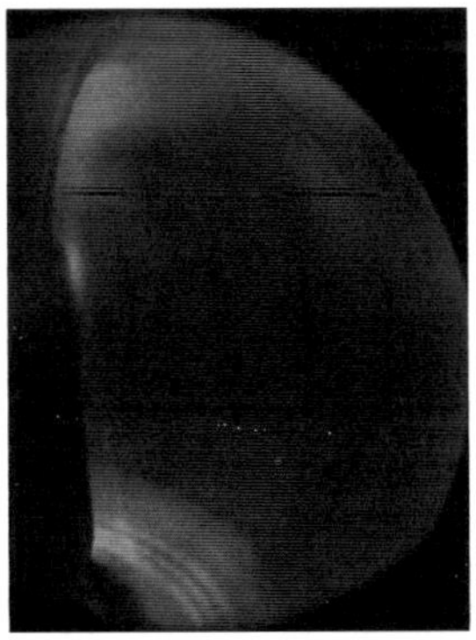

**Figure 2.** Splitting of the outer strike-point observed by a fast VIS camera.

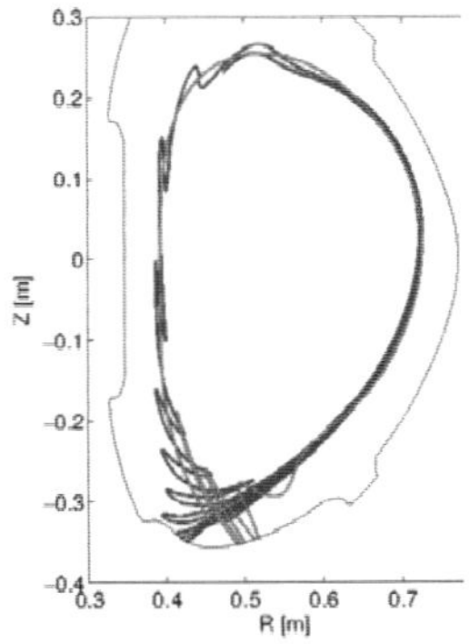


**Figure 3.** Vacuum model of the invariant manifold lobes in COMPASS.

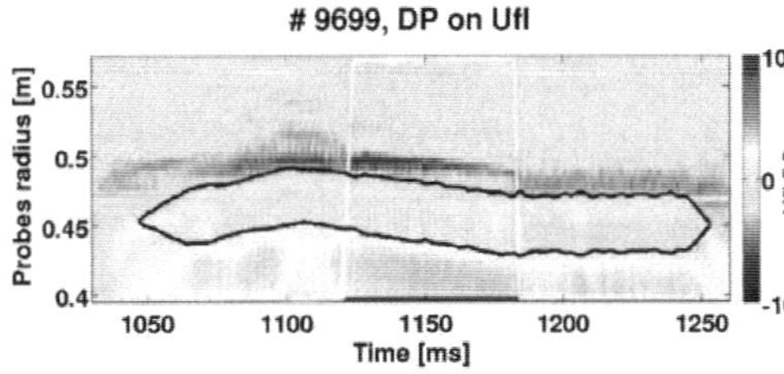


**Figure 4.** Temporal evolution of the floating potential on the divertor probes at swept strike points.

*2.2 Swept divertor probes measurements.*

To study the RMP influence on the plasma behavior during the movement of the X-point, other plasma parameters, such as the plasma potential, electron temperature and density, are of interest. For this purpose, the probes were biased with respect to the tokamak vessel by a +60 ÷ −160 V triangular voltage $U_p(t)$ swept at a frequency of 1 kHz by a KEPCO 100-4 M power supply. The probe bias $U_p(t)$ and probe current $I_p(t)$ were measured and the $I_p(U_p)$ probe characteristics were evaluated by FDPT.

As an example, figure 5 shows the results for $U_{fl}$ and the ion saturation current density $J_{sat}$ around position $R_{osp}$ of the outer strike point during L-mode discharge #9694 ($I_{pl}$ = 170 kA, $B_T$ = 1.15 T, $q_{95}$ = 3.8, $n_e^{avr}$ = 4.5×10$^{19}$ m$^{-3}$ off-midplane configuration of the RMP: +4 kA [9]) at different times during the discharge, namely, at time 1100 ms before the RMP and at 1125 ms and 1150 ms during RMP. One can see clearly pronounced splitting in the measured quantities' poloidal profiles, with positions of the minima and maxima shifted in accordance with the X-point movement, i.e. appearing at a constant value of $R$- $R_{osp}$. The splitting is observed on LFS only, though some changes are present on HFS as well. Note that the precision of the $R_{osp}$ (determined by EFIT++) is only approximately 1 cm.

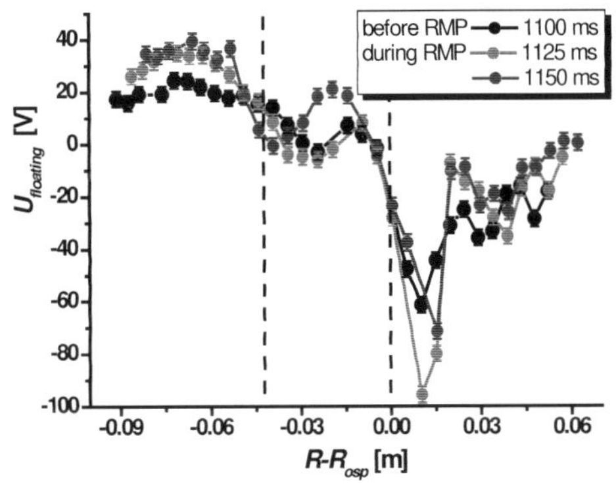


**Figure 5 a.** Poloidal profile of the floating potential before and during RMP.

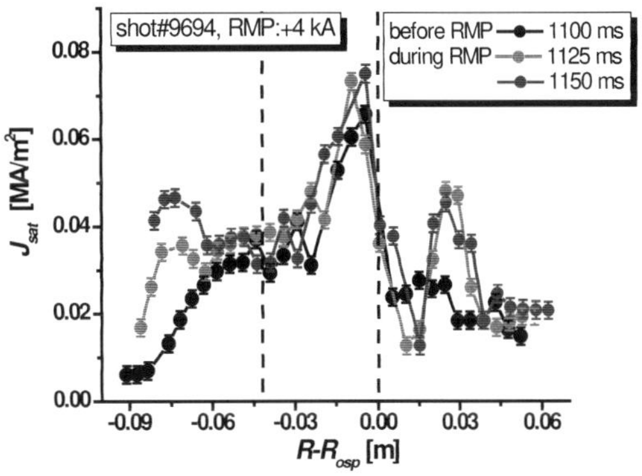


**Figure 5 b.** Poloidal profile of the ion saturation current density before and during RMP.

It is important to point out that the local minimum of $U_{fl}$ occurs at the position of the local minimum of $J_{sat}$, which is contradictory to the common expectation that the interior of the intersecting

IOP Conf. Series: Journal of Physics: Conf. Series **982** (2018) 012001 doi:10.1088/1742-6596/982/1/012001

lobes should appear as negative peak of $U_{fl}$ [14] as well as positive peak of the particle flux $J_{sat}$. This can be attributed to the SOL conditions in this COMPASS discharge being different from typical conditions in RMP experiments at larger devices, e.g. a high recycling divertor regime.

The profile of $U_{fl}$ is influenced by the plasma potential $U_{pl}$ and by the shape of EEDF (mostly by its high-energy tail); both can be obtained by the FDPT. In figure 6a we can see that the poloidal variation of $U_{pl}$ can explain most of the variation of the $U_{fl}$ except the large negative peak at the distance $R - R_{osp} = 0.01$ m.

The EEDFs obtained deviate from Maxwellian and can be approximated by bi-Maxwellian EEDFs with a low-energy (4-6 eV) electron fraction (triangles in figure 6 b) and a high-energy (11-35 eV) electron group (squares). Similar results were obtained in our previous investigations [11]; the presence of the low-energy group is attributed to the ionization process as confirmed by a model [12].

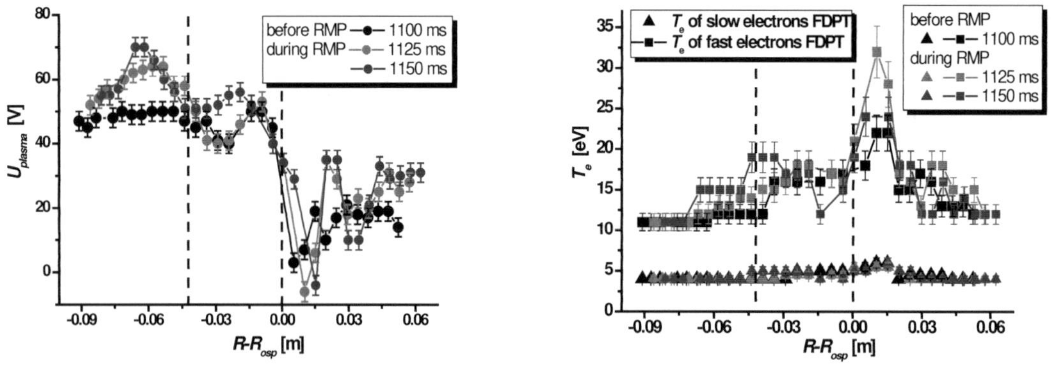


a) b)

**Figure 6.** Poloidal profile of the plasma potential **a)** and electron temperatures **b)** for discharge #9694

The splitting due to the RMP field is weakly pronounced on the electron temperature profiles as expected from the comparison between the profiles of $U_{fl}$ and $U_{pl}$. The temperature of the high-energy electron group $T_e^h$ increases by about 10 eV at the distance 0.01 m at time 1125 ms which explains the large negative peak of $U_{fl}$. The peak is less pronounced in the profile at 1150 ms, where the maximum is located in between two probes (the strike points were shifted towards the HFS).

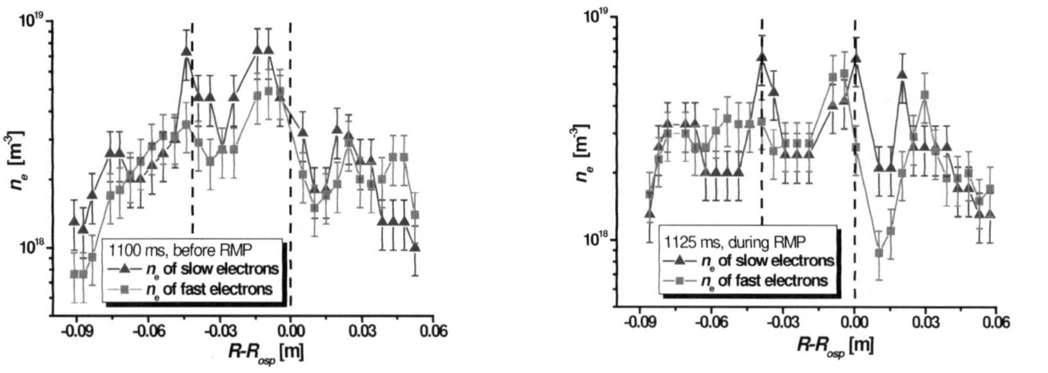


a) b)

**Figure 7.** Poloidal profile of the electron density **a)** before and **b)** during RMP, discharge #9694.

Evaluation of densities of both electron groups by the FDPT is hampered by a relatively larger error but we can distinguish the qualitative change between the profiles during RMP (figure 7 b) and before RMP (figure 7 a). The density of the high-energy electrons decreases after application of RMP by a factor of 2 at the distance 0.01 m which is clearly correlated with the increase of $T_e^h$. This is consistent with the drop of $J_{sat}$ at this position after application of RMP (figure 5 b). It is interesting to note that a local minimum of the density of both electron groups is present at this position also before the RMP pulse, although more difficult to distinguish in $J_{sat}$ due to the linear scale.

The parallel power flux density can be calculated from the effective temperature $T_e^{eff}$ which takes into account the temperatures and densities of both electron groups [12]:

$$Q_{\|} \cong \gamma \cdot T_e^{eff} J_{sat} \quad (1)$$

Here $\gamma$ is the sheath heat transition coefficient and its value is in the range of 11-24 depending on the ratio $T_e^h / T_e^{eff}$ as proposed in [12] for the case of a bi-Maxwellian EEDF. The ion temperature $T_i$ is less affected by atomic processes and thus typically higher than the electron temperature in the SOL; $T_i = T_e^h$ is assumed here for the power flux calculation. The ratio $T_e^h / T_e^{eff}$ is presented in figure 8, the resulting parallel power flux in figure 9.

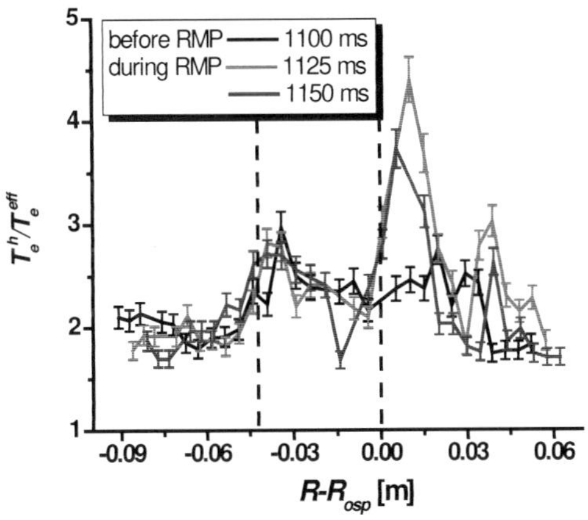


**Figure 8.** The ratio between the high-energy electron group temperature and the effective temperature.

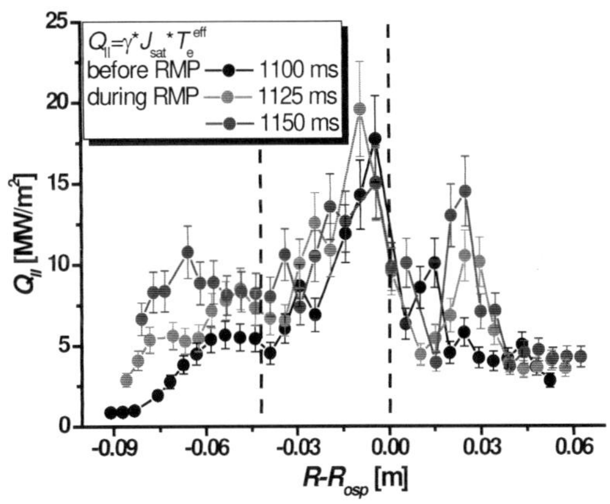


**Figure 9.** Poloidal distribution of the parallel power flux density $Q_{\|}$ .

A large effect of RMP on the $T_e^h / T_e^{eff}$ ratio is observed at the distance 0.01 m, where the temperature of the high-energy electrons increases (figure 6 b) but their density decreases at the same time (figure 7 b) and thus the effective temperature remains the same. As expected, the negative peak of $U_{fl}$ observed at this position is only related to the high-energy group and does not correlate with changes of the effective electron temperature. Therefore, the profile of the power flux density is dominated by the profile of $J_{sat}$ and we can observe clear secondary maxima which mimic those on figure 5 b.

In experiments with negative polarity of the applied RMP field, reversal of the splitting pattern in the LFS divertor profiles is observed – the maxima appear at the locations of former minima and vice versa – but the relations between the profiles of individual plasma parameters remain analogical. The splitting was observed for both the on+off-midplane and off-midplane RMP configuration [9] with very similar results.

## 3. Conclusions

The effect of RMP on the L-mode plasma parameters in divertor region of the COMPASS tokamak was studied using the array of 39 Langmuir probes embedded into the divertor target. The probe *IV* characteristics were processed by the first-derivative probe technique to obtain the plasma potential and the EEDF which was approximated by a bi-Maxwellian EEDF with a low-energy (4-6 eV) fraction and a high-energy (11-35 eV) one, the both factions having similar electron density.

Clear splitting was observed during the RMP pulse in the LFS scrape-off-layer profiles of the floating potential $U_{fl}$ and the ion saturation current density $J_{sat}$; these two quantities were obtained both by direct continuous measurement and by evaluation of the *IV* characteristics of probes with swept bias. Surprisingly, we observe that the negative peak of $U_{fl}$ induced by RMP spatially overlaps with the local minimum of $J_{sat}$ (and $n_e$) rather than with its local maximum. On one hand this is contradictory to the common expectation that the interior of the intersecting lobes should appear as negative peak of $U_{fl}$ [14] as well as positive peak of the particle flux $J_{sat}$. On the other hand, the formation of divertor profiles depends to a large extent on the conditions in the SOL and particularly at the divertor target, which can vary substantially between different experiments and devices (such as high-recycling vs. low-recycling divertor).

The common interpretation that decreased $U_{fl}$ should be attributed to increased electron temperature does not hold in all cases. In this work we show that most of the spatial variation of $U_{fl}$ is caused by the variation of $U_{pl}$ except for the largest negative peak induced by RMP. At the position of this negative $U_{fl}$ peak, we observe an increase only in the temperature of the high-energy electron fraction but the relative share of this fraction decreases at the same time, and the effective temperature of the whole EEDF is thus not affected.

The profile of the parallel power flux density is influenced mainly by the profile of $J_{sat}$ and therefore we can observe formation of secondary maxima at the LFS during the RMP pulse.

**Acknowledgements**
This research has been partially supported by International Atomic Energy Agency (IAEA) Research Contract No 17125/R0, R1 and R2 as a part of the IAEA CRP F13014 on "Utilisation of a Network of Small Magnetic Confinement Fusion Devices for Mainstream Fusion Research" 2014 IAEA Joint Experiment, by the Joint Research Project between the Institute of Plasma Physics of the CAS and the Institute of Electronics BAS BG, by the Czech Science Foundation grant No. 16-24724S and by MEYS project # LM2015045. This work has been also carried out within the framework of the EUROfusion Consortium and has received funding from the EURATOM research and training programme 2014-2018 under grant agreement No 633053 and the Co-fund by MEYS project number 8D15001. The views and opinions expressed herein do not necessarily reflect those of the European Commission.

**References**

[1] Loarte A et al 2014 *Nucl. Fusion* **54** 033007
[2] Liang Y et al 2007 *Phys. Rev. Lett.* **98** 265004
[3] Evans T E et al 2008 *Nucl. Fusion* **48** 024002
[4] Kirk A et al 2015 *Nucl. Fusion* **55** 043011
[5] Nardon E et al 2011 *J. Nucl. Mater.* **415** S914–S917
[6] Conway G D, Fietz S, Müller H W, Lunt T, Simon P, Suttrop W, Maraschek M, Happel T, Viezzer E and the ASDEX Upgrade Team 2015 *Plasma Phys. Control. Fusion* **57** 014035
[7] Panek R et al 2016 *Plasma Phys. Control. Fusion* **58** 014015
[8] Cahyna P, Panek R, Fuchs V, Krlin L, Becoulet M, Huysmans G and Nardon E 2009 *Nucl. Fusion* **49** 055024
[9] Markovic T et al 2016 *Nucl. Fusion* **56** 092010
[10] Popov Tsv K, Ivanova P I, Stockel J and Dejarnac R 2009 *Plasma Phys. Control. Fusion* **51** 065014
[11] Dimitrova M, Dejarnac R, Popov Tsv K, Ivanova P, Vasileva E, Kovačič J, Stöckel J, Havlicek J, Janky F and Panek R 2014 *Contrib. Plasma Phys.* **54** No. 3, 255
[12] Popov Tsv K et al 2015 *Plasma Phys. Control. Fusion* **57** 115011
[13] Cahyna P, Nardon E and JET EFDA Contributors 2011 *J. Nucl. Mater.* **415** S927–S931
[14] Watkins J G, Evans T E, Jakubowski M, Moyer R A, Schmitz O, Wingen A, Fenstermacher M E, Joseph I, Lasnier C J and Rudakov D L 2009 *J. Nucl. Mater.* **390–391** 839–842

IOP Conf. Series: Journal of Physics: Conf. Series **982** (2018) 012002 doi:10.1088/1742-6596/982/1/012002

# Electron energy distribution function in the divertor region of the COMPASS tokamak during neutral beam injection heating

**E Hasan[1,2], M Dimitrova[2,3,5], J Havlicek[3], K Mitošinková[3,4], J Stöckel[3], J Varju[3], Tsv K Popov[1], M Komm[3], R Dejarnac[3], P Hacek[3,4], R Panek[3] and the COMPASS team**

[1]Faculty of Physics, St. Kl. Ohridski University of Sofia, 5 J. Bourchier Blvd., 1164 Sofia, Bulgaria
[2]Acad. E. Djakov Institute of Electronics, Bulgarian Academy of Sciences, 72 Tsarigradsko Chaussee, 1784 Sofia, Bulgaria
[3]Institute of Plasma Physics, The Czech Academy of Sciences, Za Slovankou 3, 182 00 Prague 8, Czech Republic
[4]Faculty of Mathematics and Physics, Charles University in Prague, Prague, Czech Republic

E-mail: dimitrova@ipp.cas.cz

**Abstract**. This paper presents the results from swept probe measurements in the divertor region of the COMPASS tokamak in D-shaped, L-mode discharges, with toroidal magnetic field $B_T = 1.15$ T, plasma current $I_p = 180$ kA and line-average electron densities varying from 2 to $8\times10^{19}$ m$^{-3}$. Using neutral beam injection heating, the electron energy distribution function is studied before and during the application of the beam.
The current-voltage characteristics data are processed using the first-derivative probe technique. This technique allows one to evaluate the plasma potential and the real electron energy distribution function (respectively, the electron temperatures and densities).
At the low average electron density of $2\times10^{19}$ m$^{-3}$, the electron energy distribution function is bi-Maxwellian with a low-energy electron population with temperatures 4-6 eV and a high-energy electron group 12-25 eV. As the line-average electron density is increased, the electron temperatures decrease. At line-average electron densities above $7\times10^{19}$ m$^{-3}$, the electron energy distribution function is found to be Maxwellian with a temperature of 6-8.5 eV.
The effect of the neutral beam injection heating power in the divertor region is also studied.

## 1. Introduction

The probe technique is frequently applied to diagnose the magnetized high-temperature strongly turbulent fusion edge plasmas. The knowledge of the real electron energy distribution function (EEDF) is of great importance in understanding the physics underlying the processes occurring in the magnetized plasma.

[5] To whom any correspondence should be addressed.

This paper presents the results from swept probe measurements in the divertor region of the COMPASS tokamak [1, 2] in D-shaped, L-mode additionally heated deuterium plasmas at $I_p = 180$ kA, toroidal magnetic field $B_T = 1.15$ T and line-average electron densities varying from $n_e^{avr} = 2\times10^{19}$ m$^{-3}$ to $8\times10^{19}$ m$^{-3}$. The neutral beam injector [3] injects the deuterium beam tangentially into the tokamak vessel in the same direction as the plasma current (co-injection). The energy of the beam is kept at $E_{NBI} = 40$ keV. The beam power $P_{NBI}$ is varied by changing the value of the ion current extracted from the ion source $I_{NBI}$ from ~200 kW ($I_{beam} = 6$ A) to ~320 kW ($I_{beam} = 10$ A). The plasma potential and the EEDF as obtained by using the first-derivative probe technique (FDPT) [4-7] are derived before and during the application of the beam power.

The dependence of the plasma potential and the EEDF on the neutral beam injection (NBI) heating power is studied.

## 2. Langmuir probe measurements in the divertor region of the COMPASS tokamak

The divertor probe system in the COMPASS tokamak consists of 39 single graphite Langmuir probes poloidally embedded in the divertor tiles providing data with a 5 mm spatial resolution [5]. The probes are biased with respect to the tokamak chamber wall by a voltage $U_p$ swept as triangular waveform at a frequency of 1 kHz supplied by a KEPCO 100-4M power supply. The probe bias and probe current are recorded by the COMPASS data acquisition system at a sampling frequency of 1 MHz. The current-voltage (*IV*) probe characteristics are constructed using the data recorded at different times before and during the NBI heating.

Further, the FDPT is applied to process the *IV* characteristics in order to obtain the plasma potential and the EEDF (respectively, the electron temperatures and densities) [4-7]. Figure 1 shows as an example two experimental *IV* characteristics from probe #10 (radial position $R = 0.4326$ m) before NBI heating at time $t = 1080$ ms, beginning of the discharge is at 970 ms. The *IV*s at $n_e^{avr} = 2\times10^{19}$ m$^{-3}$ and $n_e^{avr} = 7\times10^{19}$ m$^{-3}$ are presented by blue and black lines. Figure 2 presents the first derivative of the same *IV* curves. In both figures by red and green lines the comparison with the model curves are presented. Using this comparison [8], one can evaluate the plasma potentials, which are 48 V and 26 V, respectively.

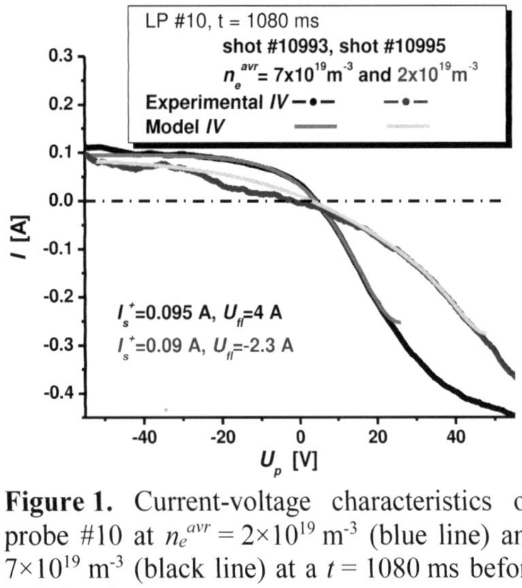


**Figure 1.** Current-voltage characteristics on probe #10 at $n_e^{avr} = 2\times10^{19}$ m$^{-3}$ (blue line) and $7\times10^{19}$ m$^{-3}$ (black line) at a $t = 1080$ ms before NBI heating.

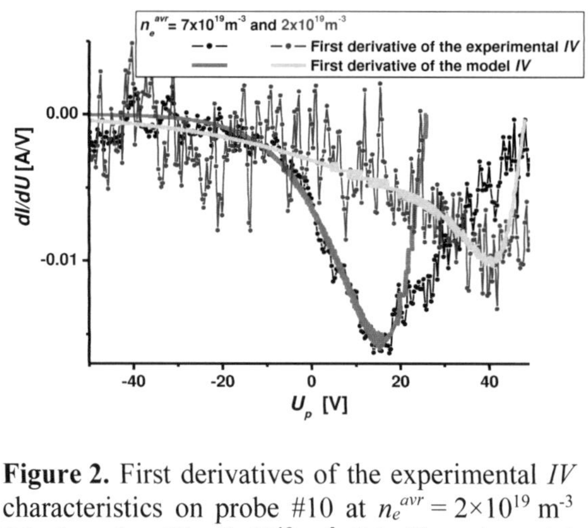


**Figure 2.** First derivatives of the experimental *IV* characteristics on probe #10 at $n_e^{avr} = 2\times10^{19}$ m$^{-3}$ (blue) and $n_e^{avr} = 7\times10^{19}$ m$^{-3}$ (black) and model curves, green and red curves respectively.

The corresponding electron energy probability functions (EEPF) [7] for both cases are plotted in figure 3. The EEPF provides the same information about the electron component as the EEDF and is frequently used to represent the probe data measured.

The EEPF, presented in a semi-log scale, allows a quick visualization of a departure of the measured EEPF from the Maxwellian distribution, which is a straight line in this representation [7]. Using the slope of the EEPF, one can determine the electron temperature.

It is seen that at $n_e^{avr} = 7\times10^{19}$ m$^{-3}$ EEPF can be approximated by a straight line. In this case we have a Maxwellian distribution with $T_e = (7.5\pm0.7)$ eV. In the case of $n_e^{avr} = 2\times10^{19}$ m$^{-3}$, the electron distribution differs from the Maxwellian, but can be approximated by a bi-Maxwellian one, i.e., by a sum of two distributions with two different temperatures: 4.5 eV and 18 eV. This is in agreement with our previous studies [5, 9].

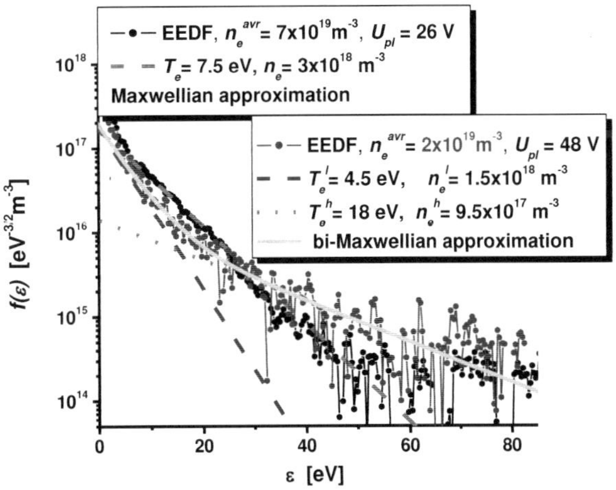


**Figure 3.** Experimental EEPF at $7\times10^{19}$ m$^{-3}$ (black dots), a Maxwellian distribution (red) at $2\times10^{19}$ m$^{-3}$ (blue dots) and a bi-Maxwellian distribution (green) - sum of the model ones for two electron temperatures for LP#10, $R = 0.4326$ m.

## 3. Results and discussion

We present the main plasma parameters estimated using the EEDF as derived from the *IV* characteristics before and during the application of the NBI heating. The results of determining the plasma potential's poloidal distribution and the electron temperatures are shown in figures 4 before a) and c), during NBI heating (b) and d)), when the beam current is $I_{beam} = 6$ A. The different values of the line-average electron densities are indicated by different colours. The positions of the strike points (presented as regions) found from the reconstruction of the magnetic surfaces by the Equilibrium FITing code (EFIT) are also shown in all figures below.

It is seen that as the line-average electron density increases, the amplitude of the $U_{pl}$ decreases. The difference in the values is strongly pronounced on the high-field side (HFS), where for $n_e^{avr} = 2\times10^{19}$ m$^{-3}$ the plasma potential is $U_{pl} = 70$ V, while for $n_e^{avr} = 8\times10^{19}$ m$^{-3}$ it is around $U_{pl} = 30$ V (figure 4 a)).

When the NBI heating is applied, the influence of the beam in the divertor is not too strong. A small increase in the range of a few volts for the $U_{pl}$ is seen, better manifested at higher line-average electron densities. On the low field side (LFS) the change is bigger again for the higher line-average electron densities, but in a negative direction, the minima deepening (figure 4 b)). Concerning the election temperature when the line-average electron densities are varied, one can see that at $n_e^{avr} = 2\times10^{19}$ m$^{-3}$ the EEDF is bi-Maxwellian. When the density is increased to $n_e^{avr} = 6\times10^{19}$ m$^{-3}$, one finds that on the HFS and in the private flux region (PFR) the EEDF is Maxwellian. On the LFS it is again bi-Maxwellian. Above $n_e^{avr} = 7\times10^{19}$ m$^{-3}$, the EEDF is Maxwellian with a temperature 6-8.5 eV. The dependence of the NBI heating with $I_{beam} = 6$A is really weak: the electron temperature increases by few eV. The influence of the two-temperature EEDF on the parallel heat flux density, as well as a discussion on the origin of the bi-Maxwellian EEDF, are presented in detail in our recent publication [6].

The electron density is calculated using the expression for normalization of the EEDF [4]. In figure 5, an example is given of the poloidal distribution of the electron densities as determined using the FDPT. The case of a bi-Maxwellian EEDF corresponds to the low line-average density $n_e^{avr} = 2\times10^{19}$ m$^{-3}$ (figure 5 a). The low-temperature electron group (blue triangles, $T_e^l = 4\pm0.4$ eV) has a density similar or higher than the high-temperature one (red squares, $T_e^h = 12\text{-}25\pm0.2$ eV). At $n_e^{avr} = 6\times10^{19}$ m$^{-3}$ (figure 5 b) on the HFS and in PFR, the EEDF is Maxwellian. On the LFS it is bi-Maxwellian. This peculiarity is not affected by the NBI heating.

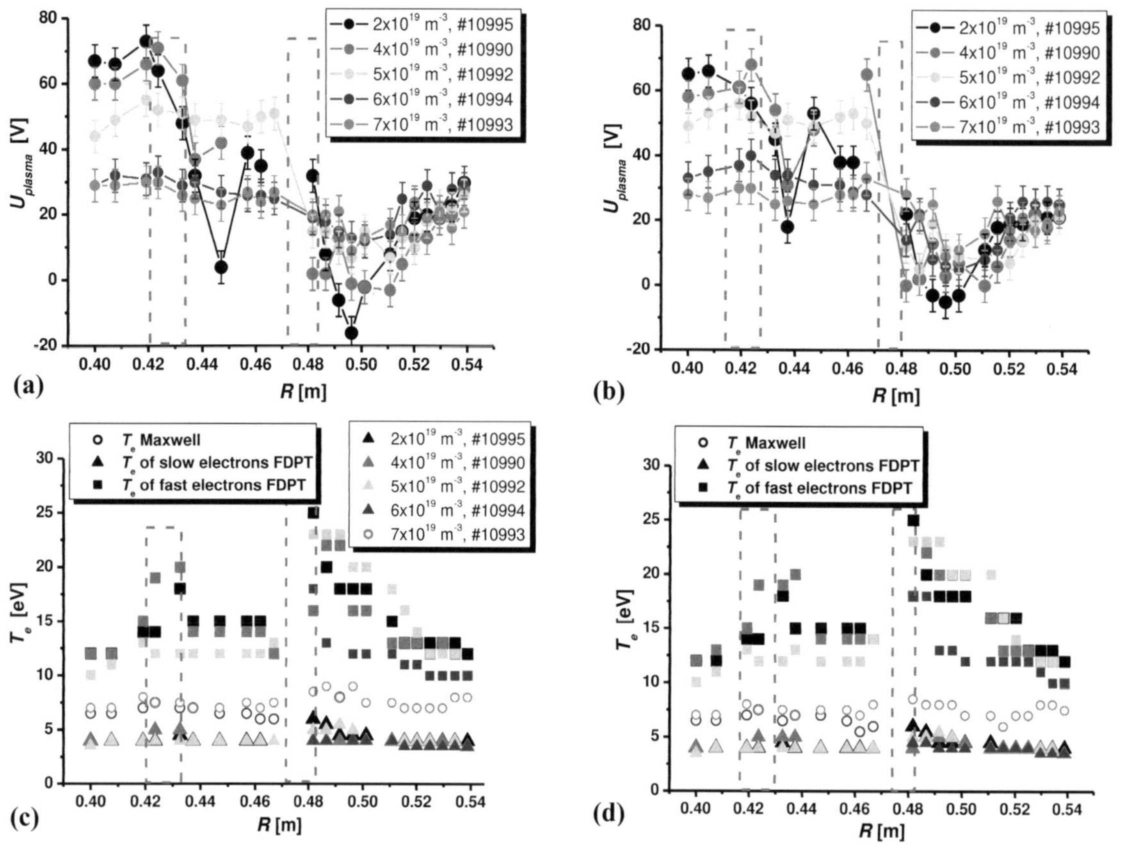


**Figure 4.** Poloidal distribution of the plasma potential **a)** before and **b)** during $I_{beam}$ =6 A NBI heating and the electron temperatures **c)** before and **d)** during NBI heating.

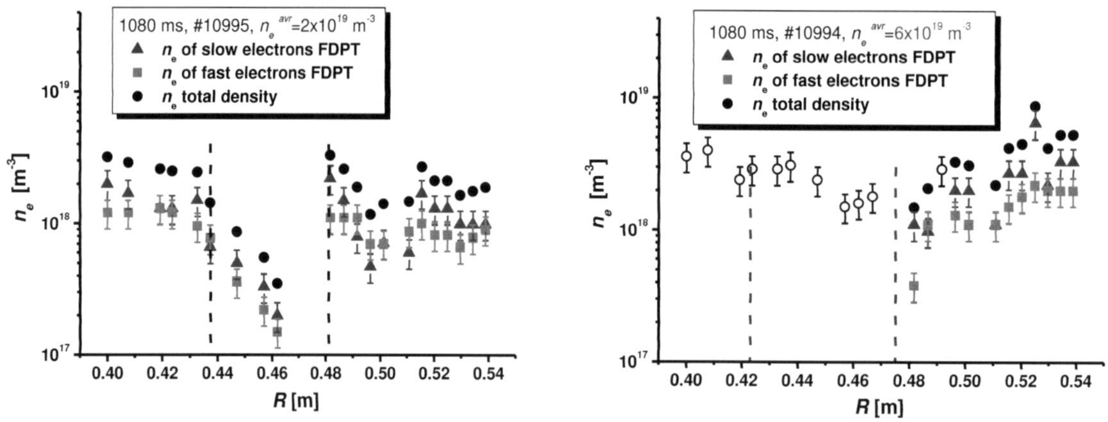


**Figure 5.** Poloidal distribution of the electron density at **a)** $n_e^{avr} = 2\times10^{19}$ m$^{-3}$ and **b)** $n_e^{avr} = 6\times10^{19}$ m$^{-3}$ before NBI heating.

The weak effect at $I_{beam}$ = 6 A can be explained by the fact that at values below 10 A the divergence of the neutral beam is rather high and a significant fraction of the neutral beam power is lost in a relatively narrow beam duct connecting NBI with the tokamak vessel.

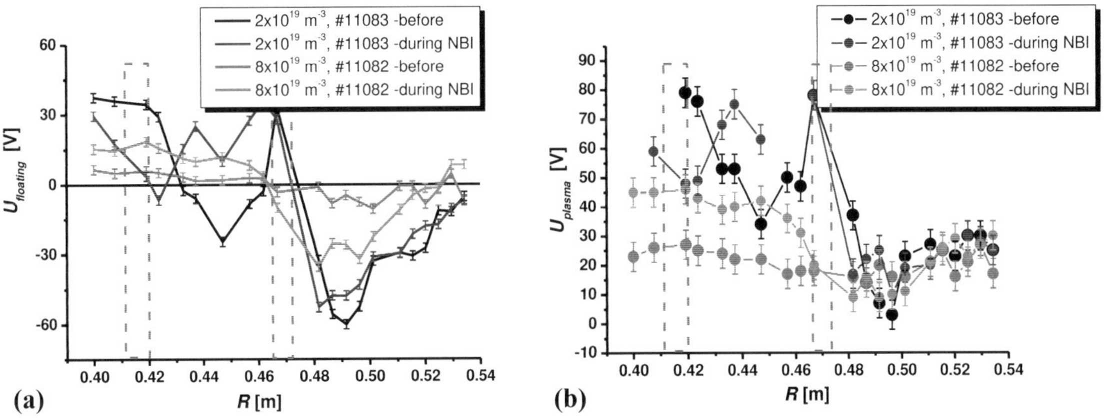


The last figure demonstrates the effect of the NBI current's higher value ($I_{beam}$ = 10 A) for $n_e^{avr} = 2\times10^{19}$ m$^{-3}$ and $8\times10^{19}$ m$^{-3}$. Figure 6 a) shows the poloidal profile of the floating potential. At higher NBI heating, the influence in the divertor region is more pronounced when the line-average density is higher. The $U_{fl}$ amplitude is larger (magenta line), more positive in the HFS and PFR and more negative in the LFS. For $n_e^{avr} = 2\times10^{19}$ m$^{-3}$, the change is distinct when the $U_{fl}$ decreases in the HFS and increases in the PFR. On the LFS, the $U_{fl}$ values before and during NBI are almost the same. The behavior of the plasma potential profile (figure 6 b)) is similar in the HFS and the PFR; however, in the LFS the values of $U_{pl}$ do not change significantly without any dependence on the density and the NBI heating power. The poloidal profile of the electron temperatures is presented in figure 6 c). It is seen that at a lower line-average density, the temperatures remain almost the same. For $n_e^{avr} = 8\times10^{19}$ m$^{-3}$ before the additional heating, the EEDF is Maxwellian (red empty dots). During the heating by a higher power, which increases the electron temperature, a bi-Maxwellian distribution appears (magenta symbols).

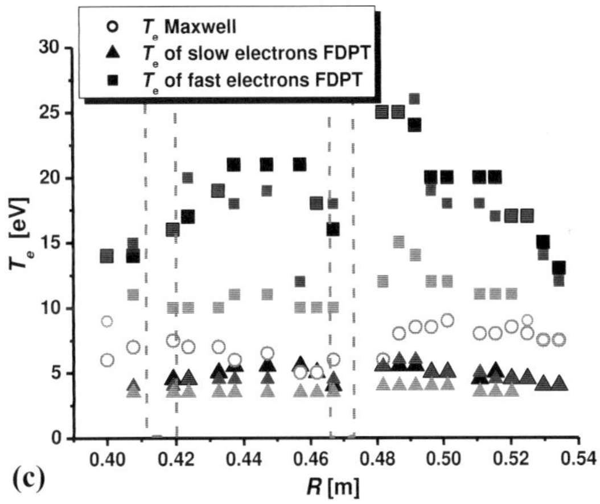


**Figure 6.** Poloidal distribution of the **a)** floating potential, **b)** plasma potential and **c)** the electron temperature for two values of the line-average density, before and during NBI heating.

## 4. Conclusions

This paper reports the results from swept probe measurements in the divertor region of the COMPASS tokamak with a plasma current of $I_p$ = 180 kA, toroidal magnetic field $B_T$ = 1.15 T and line-average electron densities varying from 2 to $8\times10^{19}$ m$^{-3}$. Using neutral beam injection heating, the electron energy distribution function is studied before and during the application of the beam.

The EEDF and the other plasma parameters in the divertor of the COMPASS tokamak are determined by the first-derivative probe technique.

At a low line-average electron density ($2\times10^{19}$ m$^{-3}$), the EEDF is bi-Maxwellian with a low-energy electron population with temperatures 4-6 eV and a high-energy electron group (12-25 eV). Raising the line-average electron density leads to a decrease in the electron temperatures. Above $n_e^{avr} = 7\times10^{19}$ m$^{-3}$, the EEDF is found to be Maxwellian with a temperature 6-8.5 eV.

The effect of NBI heating in the divertor region is more pronounced at values higher than $I_{beam}$ = 10 A.

**Acknowledgements**
This research has been partially supported by the Joint Research Project between the Institute of Plasma Physics of the CAS and the Institute of Electronics BAS BG, by MEYS project # LM2015045 and by the Co-fund under MEYS project # 8D15001. This work has been carried out within the framework of the EUROfusion Consortium and has received funding from the EURATOM research and training programme 2014-2018 under grant agreement No 633053. The views and opinions expressed herein do not necessarily reflect those of the European Commission.

**References**

[1] Panek R, et al 2016 *Plasma Phys. Control. Fusion* **58** 014015
[2] Weinzettl V, et al 2011 *Fusion Engineering and Design* **86** 1227–1231
[3] Deichuli P et al 2012 *Rev. Sci. Instrum.* **83** 02B114
[4] Popov Tsv K, Ivanova P I, Stockel J and Dejarnac R 2009 *Plasma Phys. Control. Fusion* **51** 065014
[5] Dimitrova M, Dejarnac R, Popov Tsv K, Ivanova P, Vasileva E, Kovačič J, Stöckel J, Havlicek J, Janky F and Panek R 2014 *Contrib. Plasma Phys.* **54** No. 3 255–60
[6] Popov Tsv K, et al 2015 *Plasma Phys. Control. Fusion* **57** 115011
[7] Popov Tsv K et al 2016 *Plasma Sources Sci. Technol.* **25** 033001
[8] Popov Tsv K et al 2014 *Contrib. Plasma Phys.* **54** No. 3 267272
[9] Ivanova P, Dimitrova M, Vasileva E, Popov Tsv K, Dejarnac R, Stöckel J, Imríšek M, Hacek P and Panek R 2016 *J. Phys.: Conf. Series* **768** 012003

IOP Conf. Series: Journal of Physics: Conf. Series **982** (2018) 012003 doi:10.1088/1742-6596/982/1/012003

# Digital filter polychromator for Thomson scattering applications

**V Solokha[1,2], G Kurskiev[2], E Mukhin[2], S Tolstyakov[2], N Babinov[1,2], A Bazhenov[2], I Bukreev[2], A Dmitriev[2], M Kochergin[2], A Koval[2], A Litvinov[2], S Masyukevich[3], A Razdobarin[2], D Samsonov[2], V Semenov[2], V Solovey[3], P Chernakov[3], Al Chernakov[3], An Chernakov[3]**

[1]Peter the Great St. Petersburg Polytechnic University, Saint-Petersburg, Russia
[2]Ioffe Institute, Saint-Petersburg, Russia
[3]SpectralTech Ltd., Saint-Petersburg, Russia

E-mail: v.solokha@spbstu.ru

**Abstract**. Incoherent Thomson scattering diagnostics (TS) is a proven technique capable of reliable and robust instantaneous measurement of electron temperature ($T_e$) and density ($n_e$) local values in wide area of plasma physics experiments: from hall-effect thrusters to tokamaks and stellarators. The TS cross section is very low (~ $6.7\times10^{-30}$ m$^2$), and the corresponding TS signals, measured in fusion experiments, are usually of ~$10^{-15}$ of incident power. This paper represents 6 (7) channel filter polychromator equipped with avalanche photodiodes and low-noise preamplifiers. The incorporated ADC system (5 GS/s, 12 bit) provides digital optical output preventing acquisition system from electromagnetic interferences. The calibration techniques and $T_e$, $n_e$ with corresponding errors measured in Globus-M plasma are given for the digital polychromator test-bench.

## 1. Introduction

Incoherent Thomson scattering (TS) is one of the key diagnostics for measurements of electron temperature and density profiles in modern tokamaks. Study of such processes as L-H transitions, Internal Transport Barriers, Edge Localized Modes instability together with ability to use the diagnostic as a sensor for real-time plasma control require reliable multiple laser and spectral analytical systems. There are several approaches to TS spectrum analyses. These are diagnostic techniques based on matrix detector (CCD camera, Image Intensifier) on the output of filter [1] and grating polychromators [2] and avalanche photodiodes (APD), which can also be used with both filter [3] and grating [4] devices. The Globus-M TS system based on multiple pulsed Nd:Glass laser (1055 nm, 2 J, 30 ns, up to 20 pulses in burst mode with repetition rate up to 3 kHz) and Divertor system based on Nd:YAG laser (1064 nm, 2 J, 3 ns, 100 Hz). The laser wavelength choice prohibits the use of matrices sensors due to low quantum efficiency in the vicinity of 1000 nm and causes the choice of silicon APDs. Interference-filter-based polychromators with the APDs [3] have a high throughput and are routinely used for TS diagnostics in fusion devices. Laser pulse of 3 ns (FWHM) and high-speed electronics developed for Divertor TS system are chosen to resolve in time TS signals and stray light reflected from the closely spaced first wall as well as for minimizing signals of plasma background and heated surface radiation. Filter polychromator described in this paper, will be used for both the Equatorial Edge and Divertor TS systems on Globus-M tokamak [5] with the following parameters:

major radius $R$ = 36 cm and minor radius $a$ = 24 cm, toroidal magnetic field $B_T$ = 0.4 T, core $n_e$ ~$5\times10^{12}$ to $2\times10^{14}$ $cm^{-3}$ and core $T_e$ from 20 eV to 1.5 keV (see figure 1). The equatorial edge and divertor TS systems, to be launched next year, should provide measurements in 9 spatial points for divertor region and 5 at the edge – near the pedestal region. The edge $T_e$ are much lower than core ones [6] and require to use more narrow spectral channels.

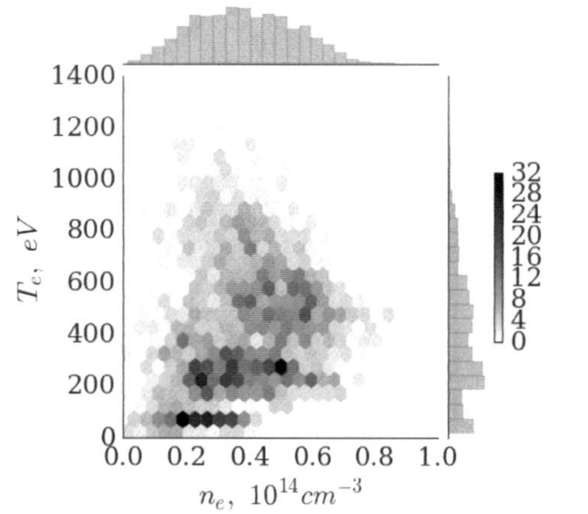


**Figure 1.** Globus-M typical $T_e$ and $n_e$ values

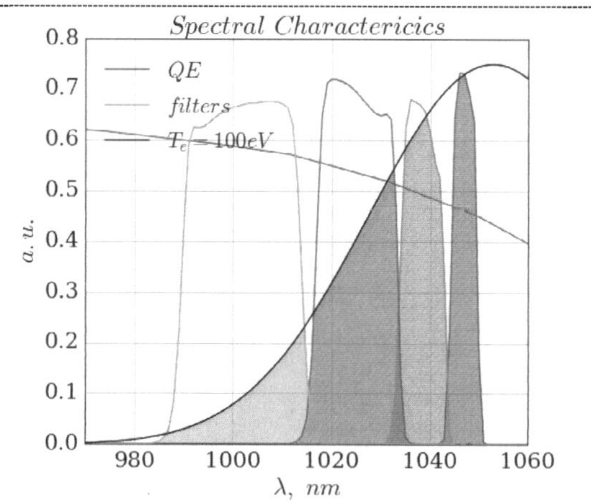


**Figure 2.** Spectral channels of edge TS polychromator, APD quantum efficiency and TS spectrum for $T_e$= 100 eV

## 2. Data handling

### 2.1. Signal processing

Principles of $T_e$ and $n_e$ measurements that underlie the TS diagnostics are quite simple: $n_e$ is determined from overall scattered light intensity and temperature comes from Doppler broadening of TS spectra (figure 2). Laser beam propagates through the plasma and scattered by free electrons. The scattered radiation is collected by fiber bundle and delivered to the high-transmission polychromators [7]. The 1.5 mm APD's are equipped with low-noise transimpedance preamplifiers and the output signal are digitized in oscilloscope mode. Such approach looks the most flexible to distinguish TS signal against stray light; to record signal prehistory for background noise estimation and to process signal with the assistance of required tools of mathematics.

The number of scattered photons in spectral channels were calculated in accordance with:

$$N_{pe} = \frac{\int_\tau U_{signal}dt}{M\cdot G\cdot R_f\cdot e\cdot R_{out}/(R_{in}+R_{out})}\ , \tag{1}$$

where $U_{signal}$ – amplified signal, $\tau$ – laser pulse duration, $M$ – APD gain, $G$ – multiplication factor of amplifier, $R_f$ – feedback resistances of transimpedance preamplifier, $e$ – electron charge, $R_{out}$ and $R_{in}$ – amplifier output and ADC input resistances.

### 2.2. Temperature and density estimation

The number of photoelectrons in i[th] channel can be written as:

$$\mathbf{N_{fe_i}} = n_e\cdot L\cdot \Omega\ \cdot E_L\cdot \frac{\lambda_0}{hc}\cdot \int_{\lambda^i_{MIN}}^{\lambda^i_{MAX}} \sigma_{TS}\cdot T^i_{over}\cdot QE^i\cdot K^i\, d\lambda\ , \tag{2}$$

where $n_e$ – electron density, $L$ – scattering length, $\Omega$ – solid angle, $E_L$ – laser pulse energy, $\lambda_0$ – incident light wavelength, $h$ – Planck constant, $c$ – speed of light, $\lambda_{MIN}$ and $\lambda_{MAX}$ – spectral channel edges, $\sigma_{TS}$ TS cross section, $T_{over}$ – overall transmission, $QE$ – APD quantum efficiency, $K$ – filter spectral characteristics. The incoherent TS spectra for the expected in Globus-M $T_e$ values (see figure 1) can be found in a number of papers [8]. Temperature estimation is obtained from minimization of $\chi^2$ function.

$$\chi^2 = \sum_i \frac{1}{\sigma^i_{N_{TS}}} \left(N^i_{EXP} - N^i_{NUM}\right)^2 \tag{3}$$

where $N_{EXP}$ – measured number of photoelectrons, $N_{NUM}$ – number of photoelectrons for the expected $T_e$ and $n_e$. Iterative minimization of $\chi^2$ function gives the most probable values of the measured $T_e$ and $n_e$.

### 2.3. *Accuracy of estimation*

Random errors in $T_e$ and $n_e$ measurement were estimated using the following equations [9]:

$$\sigma_{T_e} \approx \left[\sum\left(\frac{N_{fe_i}}{\sigma_{N_{fe_i}}}\right)^2 \cdot \left(\sum\left(\frac{N_{fe_i}}{\sigma_{N_{fe_i}}}\right)^2 \cdot \sum\left(\frac{\partial N_{fe_i}}{\partial T_{e_i}} \cdot \frac{1}{\sigma_{N_{fe_i}}}\right)^2 - \left(\sum \frac{\frac{\partial N_{fe_i}}{\partial T_e} N_{fe_i}}{\sigma_{N_{fe_i}}^2}\right)^2\right)^{-1}\right]^{0.5} \tag{4}$$

$$\sigma_{n_e} \approx \left[n_e \sum\left(\frac{\partial N_{fe_i}}{\partial T_e} \cdot \frac{1}{\sigma_{N_{fe_i}}}\right)^2 \cdot \left(\sum\left(\frac{N_{fe_i}}{\sigma_{N_{fe_i}}}\right)^2 \cdot \sum\left(\frac{\partial N_{fe_i}}{\partial T_{e_i}} \cdot \frac{1}{\sigma_{N_{fe_i}}}\right)^2 - \left(\sum \frac{\frac{\partial N_{fe_i}}{\partial T_e} N_{fe_i}}{\sigma_{N_{fe_i}}^2}\right)^2\right)^{-1}\right]^{0.5} \tag{5}$$

Here $\sigma_{N_{fe_i}}$ is the dispersion of the photoelectrons detected in $i$-th spectral channel:

$$\boldsymbol{\sigma}_{\boldsymbol{N}_{fe_i}} = \sqrt{F^i_{APD}\left(\boldsymbol{N}_{fe_i} + 2\boldsymbol{N}_{bg_i}\right) + 2\boldsymbol{N}_{AMP_i}} \quad , \tag{6}$$

where $F^i_{APD}$ is APD excess noise factor in $i$-th channel, $\mathbf{N_{fe_i}}$ and $\mathbf{N_{bg_i}}$ – the TS and background photoelectrons before amplification in $i$-th channel, $\mathbf{N_{AMP_i}}$– the amplifier noise given also as the photoelectrons before amplification.

The equation (6) can be simplified to:

$$\sigma^2_{\mathbf{N_{fe_i}}} = F^i_{APD}\mathbf{N_{fe_i}} + {\sigma^i_0}^2, \tag{7}$$

where $\sigma^i_0$ – the total background noise including noise of electronics – is estimated from the noise analysis of the signal prehistory (see figure 3). For estimation of the effective excess noise factor $F$ we used dispersion of the signal intensity shown in figure 4. The linear fitting of the measured signals gives $F \sim 4.7$ exceeding by ~17 % the value provided by Hamamatsu S11519 data sheet for APD gain $M = 100$.

### 2.4. *Electronics*

Specially designed PHEMT-based transimpedance preamplifier provides wideband, low noise amplification of APD signal. The integration time 3-5 ns were optimized for the detected TS signal duration 3–5 ns FWHM. The acquisition system based on DRS4 chip [10] provides 12-bit (4096 steps) digitizing with a ~0.25 mV amplitude resolution and ~5 GHz sampling rate. Designed option to compensate slowly variable signal (background radiation) provides the use of full ADC dynamic range for digitizing of pulsed TS signal. For 5 ns integration time the measured level of intrinsic noise is of ~0.5 mV or ~16 noise photoelectrons given to the detector input for $M = 100$.

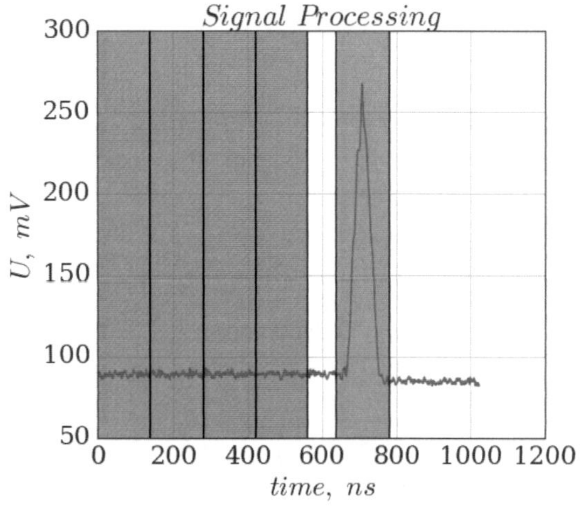


**Figure 3.** Example of oscilloscope signal and gates of signal integration. Red: range for noise calculation. Green: TS signal.

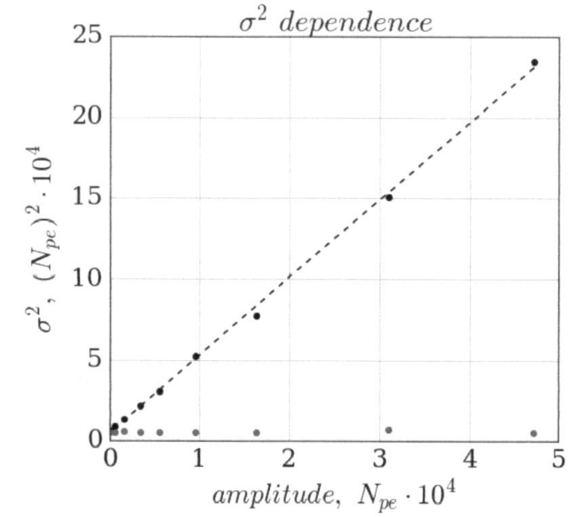


**Figure 4.** Linearity of noise dependence on signal.

## 3. Polychromator performance

It is well known that a non-uniform distribution of the spectral channel widths [3] may cover a wide spectral range by relatively small number of the channels. The current design provides $T_e$ measurements in the range of 3–600 eV by four spectral channels (see figures 5–8). These estimations were carried out for actual design of the edge and divertor Globus-M TS systems with corresponding parameters presented at the figure 1 and including background from bremsstrahlung for typical Globus-M plasma.

The relative calibration was performed using the scanning light source. It includes the tungsten band lamp and monochromator with instrumental function (FWHM) ~ 0.5 nm that is much narrower than interference filter bandwidths. Nevertheless, we have found that deconvolution may be quite important for the absolute calibration with the use of Raman scattering. Even small shift of a spectral channel edge shown by dashed curve in figure 9 may lead to a marked change in rotational Raman scattering (RRS) signal intensity if the shift includes or excludes separate spectral lines. The differences between RRS signals with and without deconvolution are presented in Table 1 for first and second spectral channels.

IOP Conf. Series: Journal of Physics: Conf. Series **982** (2018) 012003 doi:10.1088/1742-6596/982/1/012003

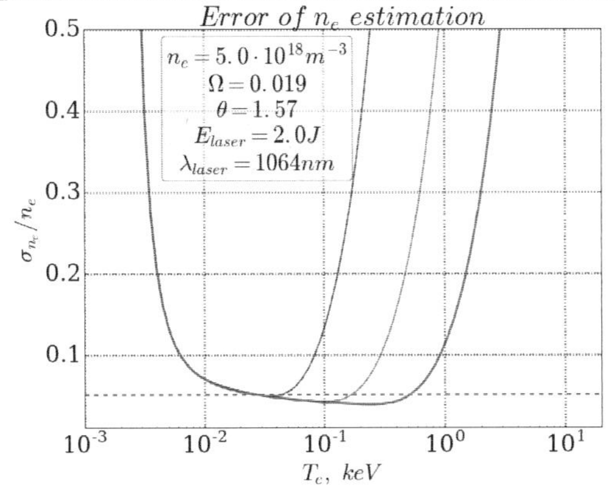


**Figure 5.** Normalized error curve of $n_e$ estimation for typical divertor plasma.

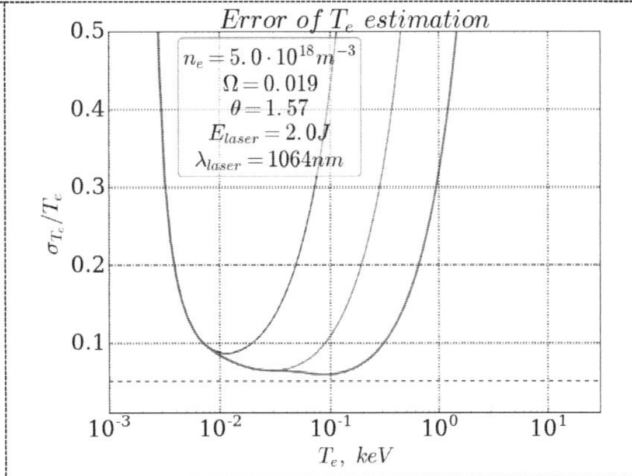


**Figure 6.** Normalized error curve of $T_e$ estimation for typical divertor plasma.

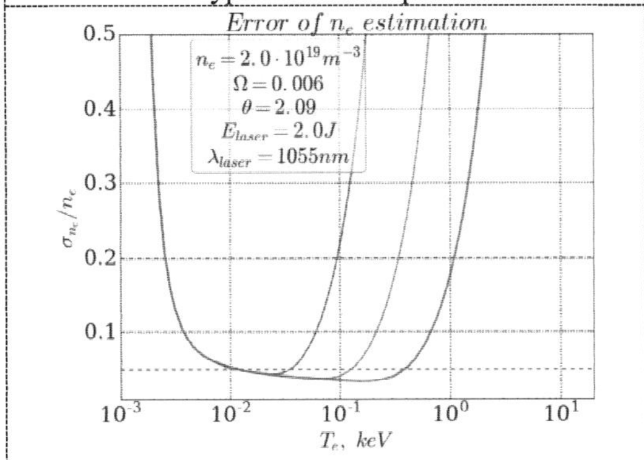


**Figure 7.** Normalized error curve of $n_e$ estimation for typical edge plasma.

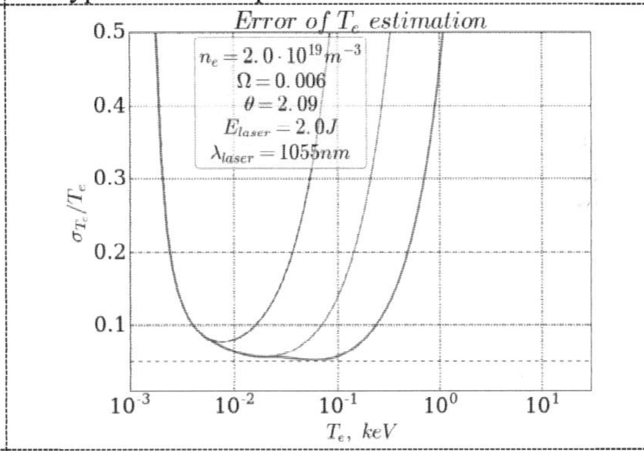


**Figure 8.** Normalized error curve of $T_e$ estimation for typical edge plasma..

**Table 1.** Calculated RRS signals at $p$ = 30 Torr.

| | RRS Signal, phe Without deconvolution | RRS Signal, phe With deconvolution | *Difference* |
|---|---|---|---|
| **Channel #1** | 939 | 934 | 0.54% |
| **Channel #2** | 258 | 250 | 3.20% |

The absolute or density calibration relies on Raman scattering for 0.03 – 10 Torr of Nitrogen. Figure 10 represents scattering signal intensity versus $N_2$ pressure. Signal for each pressure was measured 100 times that are shown by grey points. The mean values (red cross in figure 10) demonstrates good linear dependence in overall pressure range. The dispersion of the Raman scattering signals (see figure 10) gives information about overall detection accuracy for different signal values, which includes superposition of statistical signal noise, amplifier noise and other disturbing influences on the signal. The results are in good agreement with predictions made for the existing experimental layout and Raman cross section data given from [11].

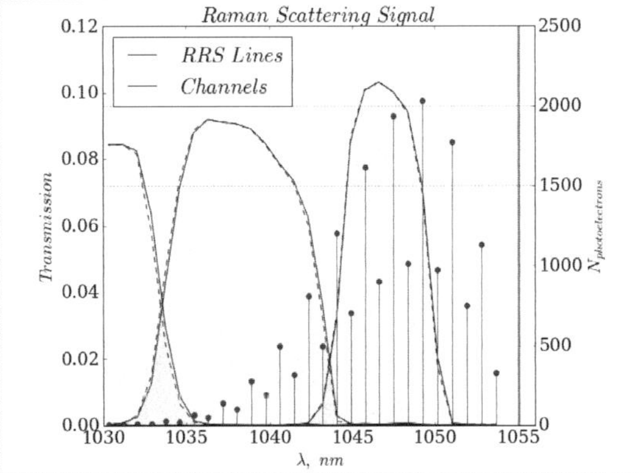


**Figure 9.** Nitrogen RRS spectrum for probing wavelength 1054 nm (blue dots), specral curves of interference filter ttransmission (black curves) and specral curves meassured with instrumental function (FWHM) ~ 0.5 nm (black dashed curves).

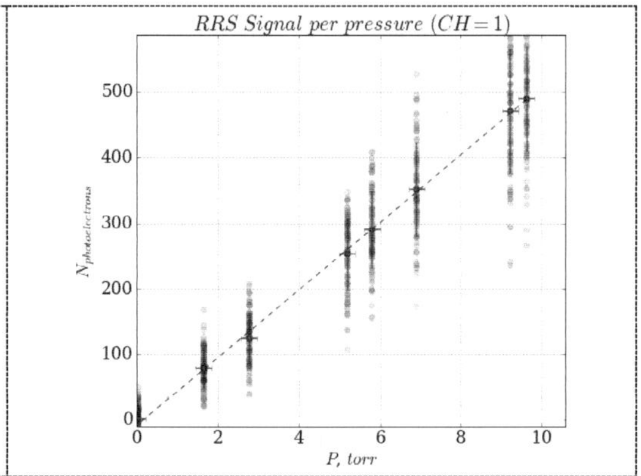


**Figure 10.** RRS signals measured by 1st spectral channel for different Nitrogen pressures.

## 4. First $T_e$ $n_e$ measurements in plasma experiment

The developed digital filter polychromator was tested during 2016 summer campaign on Globus-M. The scattered light was collected from the plasma equatorial edge ($R$ = 0.21 m), characterized by moderate electron densities $<n_e> \sim 2\times10^{19}$ m$^{-3}$ and temperatures $<T_e> \sim$ 150 eV. To test polychromators the laser 1054 nm 30 ns was used.

Figure 11 represents an example of TS signals of 1st, 2nd, 3rd and 4th spectral channels, corresponding to $n_e$ = $3\times10^{18}$ m$^{-3}$ and $T_e$ = 160 eV. The most intensive stray light manifested in the nearest to the laser wavelength – 1st channel (red mark on figure 11) and delayed by ~15 ns can be separated thanks to oscilloscopic operational mode and high sampling rate (5 GHz).

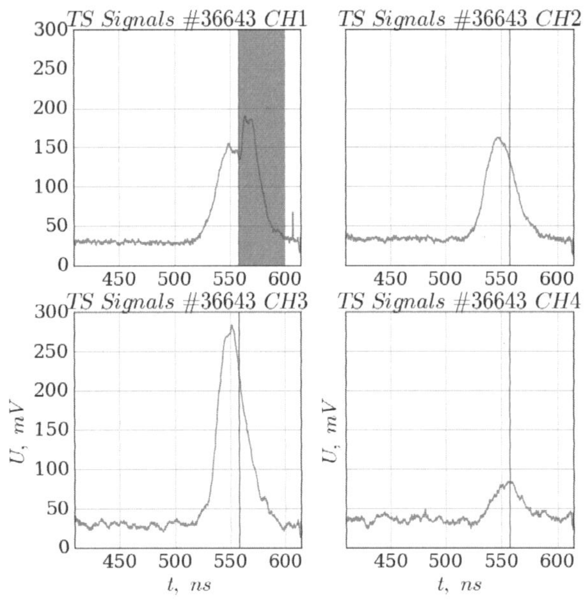


**Figure 11.** An example of TS signals of 1st, 2nd, 3rd and 4th spectral channels measured in Globus-M equatorial edge ($R$ = 0.21 m) the plasma.

## 5. Summary

The paper presents first results obtained by the digital filter polychromator, which was developed for application in Thomson scattering plasma diagnostics. The polychromator is equipped with APDs of ~1.5 mm and specially designed PHEMT-based transimpedance preamplifiers. This fast and low-noise detector unit combined with build-in ADC (12 bit, 5 GS/s) provides ~16 photoelectrons own noise level of the entire detection system recalculated on the detector input (before APD amplification) for integration time of ~5 ns. The polychromator was calibrated and tested with the use of Raman scattering (RRS) in gas Nitrogen. The dispersion of multiple RRS measurement allows performing the absolute measurement of overall system noise for different signals. This experiments shows that for signals of 100 photoelectrons and above, the noise of the RRS signal is defined mainly by statistical noise and constant excess noise factor. The measured excess noise $F\sim 4.7$ exceeding by ~17% $F = 4$ provided by Hamamatsu S11519 data sheet for APD gain $M = 100$. The estimations carried out for RRS relative calibration demonstrates the necessity to measure spectral channel characteristics with quite high spectral resolution because even small shift of a spectral channel edge may lead to a significant change in RRS signal intensity. The extremely compact size and low energy consumption of the ADC based on DRS4 chip allow combining the filter polychomator, digitizer and CPU for data processing in one stand-alone device, providing digital optical output preventing electromagnetic interferences. The first TS measurements in Globus-M demonstrate that operation in oscilloscope mode and high sampling rate (5 GHz) helps to separate in time the TS and stray light delayed by ~15 ns.

**Acknowledgments**
The first author V.Solokha acknowledges The European Fusion Network FuseNet for financial assistance of taking part in the 7th International Workshop & Summer School on Plasma Physics.

**References**

[1] A. Alfier, R. Pasqualotto Edge Thomson scattering on RFX-mod 2006 *Rev. Sci. Instr.* **77** 10E501
[2] N. L. Schoenbeck, et al. A Compact Multichannel Spectrometer for Thomson Scattering 2012 *Rev. Sci. Instr.* **83** 10E330
[3] T. N. Carlstrom, et al. A Compact, Low Cost, Seven Channel Polychromator for Thomson Scattering Measurements 1990 *Rev. Sci. Instr.* **61** 2858
[4] E. E. Mukhin et al. The ITER divertor Thomson scattering system: engineering and advanced hardware solutions, 2012 *JINST* **7** C02063
[5] G. S. Kurskiev et al. 2012 *Problems of Atomic Science and Technology* (*Series in Thermonuclear Fusion*) **35** 81–88 (in Russian)
[6] V. K. Gusev et al. Globus-M results as the basis for a compact spherical tokamak with enhanced parameters Globus-M2 2013 *Nuclear Fusion* **53** 093013
[7] https://spectraltech.ru/product/3/
[8] S. L. Prunty, A primer on the theory of Thomson scattering for high-temperature fusion plasmas 2014 *Phys. Scr.* **89**
[9] G. S. Kurskiev et al. A study of core Thomson scattering measurements in ITER using a multi-laser approach 2015 *Nucl. Fusion* **55** 053024
[10] S. Ritt, R. Dinapoli, U. Harmann, Application of the DRS chip for fast waveform digitizing, 2010 *NIMA* **623** 486-488
[11] B. P. LeBlanc, Thomson scattering density calibration by Rayleigh and rotational Raman scattering on NSTX 2008 *Rev. Sci. Instr.* **79** 10E737

IOP Conf. Series: Journal of Physics: Conf. Series **982** (2018) 012004 doi:10.1088/1742-6596/982/1/012004

# Effects of Resonant Helical Field on Toroidal Field Ripple in IR-T1 Tokamak

**B. Mahdavipour[1] , A. Salar Elahi[2*], M. Ghoranneviss[3]**
1,2,3Plasma physics Research Center, Science and Research Branch, Islamic Azad University, Tehran, Iran

salari_phy@yahoo.com

**Abstract.** The toroidal magnetic field which is created by toroidal coils has the ripple in torus space. This magnetic field ripple has an importance in plasma equilibrium and stability studies in tokamak. In this paper, we present the investigation of the interaction between the toroidal magnetic field ripple and resonant helical field (RHF). We have estimated the amplitude of toroidal field ripples without and with RHF (with different $q = m/n$) ( $m = 2$, $m = 3$, $m = 4$, $m = 5$, $m = 2$ & 3, $n = 1$) using "Comsol Multiphysics" software. The simulations show that RHF has effects on the toroidal ripples.

## 1. Introduction

Tokamak is an acronym developed from the Russian words "Toroidalnaya Kamera i Magnitnaya Katushka" which means "toroidal chamber with magnetic coils". As the name suggests, it is a magnetic confinement device with toroidal geometry [1]. In Tokamak, the outward force of the plasma pressure is balanced by the magnetic field. The magnetic field is created by toroidal coils, primary coils, and vertical coils. Toroidal coils generate the magnetic toroidal field. The plasma current produces the poloidal magnetic field. Thus, the combination of toroidal and poloidal fields give rise to magnetic field lines that have a helical trajectory around the torus. The magnetic field ripple which is made by toroidal coils has an importance in plasma equilibrium and stability studies. The magnetic field in an axisymmetric tokamak is Comprised an externally produced toroidal component,

$$B_{\phi} = B_0 \frac{1}{[1 + \left(\frac{r}{R}\right)\cos\theta\,]} \tag{1}$$

If the toroidal field is generated by N separate coils, its magnitude is no longer independent of the variable $\phi$ since the magnetic field will be slightly stronger within the plane of the coils than between the coils. This variation can be represented by the addition of a small perturbation to the toroidal field component, $\Delta B$ [2].

$$\frac{\Delta B}{B_0} = \delta(r\,,\,\Theta)\cos \mathrm{N}\,\phi \tag{2}$$

This discrepancy causes the toroidal field fluctuation named Toroidal Ripple.Low fluctuations of magnetic field cause the better confinement and plasma stability. Also, the nonlinear evolution of magnetohydrodynamical instabilities makes the plasma cool within milliseconds which create the runaway electrons,and finally confinement loss happens which named major disruption. One way to reduce the plasma disruptions in circular cross section tokamaks is to use the resonance helical filed

(RHF), which is caused by a winding coil that wrapped around the chamber torus. RHF can postpone the major disruptions [3], flatting the plasma current [4, 5], decreases the Shafaranov shifts [6], plasma displacements and amplitude of magnetic field fluctuations [5], improve the plasma confinement and increase the confinement time [4, 5]. Some experimental and theoretical studies have been done such as chaotic structure. In this paper, we present the simulation study on the effects of RHF on the toroidal field in the IR-T1 Tokamak which is the small, low beta and large aspect ratio circular cross section tokamak. Picture of the IR-T1 Tokamak, plotted by Comsol Multiphysics software in shown in figure 1.

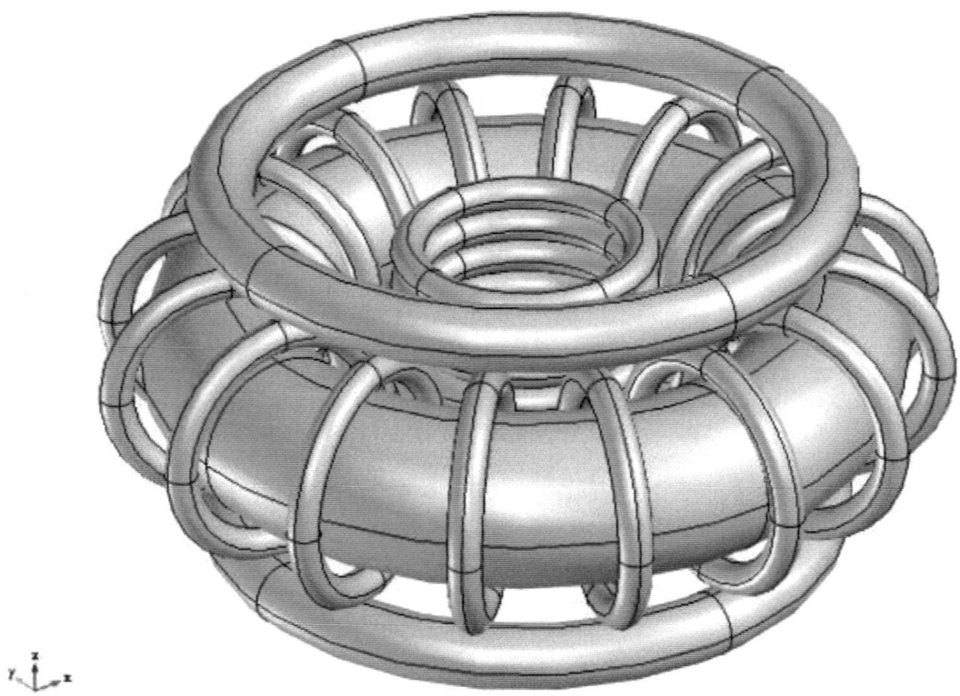

**Figure 1.** Circular cross section of the IR-T1 tokamak simulated by Comsol Multiphysics. The chamber has major and minor radius 45 cm and 12.5 cm, respectivelly. 16 O-shape TF coils and 7 primary coils are designed with the major radius of 17.5 cm and the minor radius of 2.5 cm. 2 vertical coils with the major radius of 45 and the minor radius of 5 cm. Also, we assume that there is no plasma current ($I_p$=0).

## 2. Modelling details

### 2.1. Toroidal field ripple

Toroidal coils cause the toroidal field which calculated by this equation:

$$B = \frac{\mu N I}{2\pi R} \tag{3}$$

In this equation, $R$ is the major radius of tokamak and N is the number of toroidal coils, $I$ is the current in the toroidal field coils, $\mu = 4\pi \times 10^{-7}\ \frac{\mathrm{H}}{\mathrm{m}}$.

Toroidal field lines fluctuation is called the Ripple. The main reason of this ripple is the finite number of toroidal field coils with space between them which is large enough to effects magnetic field intensity and causes the undulation. The toroidal ripple may cause the reduction of confinement and plasma stability. RHF coils are the external coils which are similar to the toroidal field coils that apply the resonance magnetic perturbation to the plasma edge. This coil effects on toroidal field ripple due to its $q$ factor, $q = m/n$ with $m$ and $n$ the toroidal and poloidal rotation around the tokamak chamber respectively [7].

### 2.2. Comsol Multiphysics

COMSOL Multiphysics® is a general purpose software platform, based on advanced numerical methods, for modeling and simulating physics-based problems. COMSOL Multiphysics, enable to account for coupled or multiphysics phenomena, also further expanding the simulation platform with dedicated physics interfaces and tools for electrical, mechanical, fluid flow, and chemical applications. Using COMSOL Multiphysics provides a significant amount of physics modeling functionality, including multiphysics ability. By adding application-specific modules, the modeling power is increased with dedicated tools for electrical, mechanical, fluid flow, and chemical applications. COMSOL Multiphysics includes a set of core physics interfaces for common physics application areas such as structural analysis, laminar flow, pressure acoustics, transport of diluted species, electrostatics, electric currents, heat transfer, and Joule heating. These are simplified versions of a selected set of physics interfaces available in the add-on modules. For arbitrary mathematics or physics simulations, where a preset physics option is not available, a set of physics interfaces is included for setting up simulations from first principles by defining the equations. Several partial differential equations (PDE) templates make it easy to model second-order linear or nonlinear systems of equations. By stacking several equations together, you can also model higher-order differential equations. These equation-based tools can further be combined with the preset physics of COMSOL Multiphysics or any of the add-on modules, allowing for fully-coupled and customized analyses. This dramatically reduces the need for writing user subroutines in order to customize equations, material properties, boundary conditions, or source terms.

## 3. Results and discussion

In this paper, we have studied the effects of RHF on toroidal field ripple using Comsol Multiphysics software. The RHF Coil with different values of $q = m/n$ modes ($n = 1$ and $m = 2, 3, 4, 5, 2\&3$) was designed and simulated. Results are plotted in figure 2.

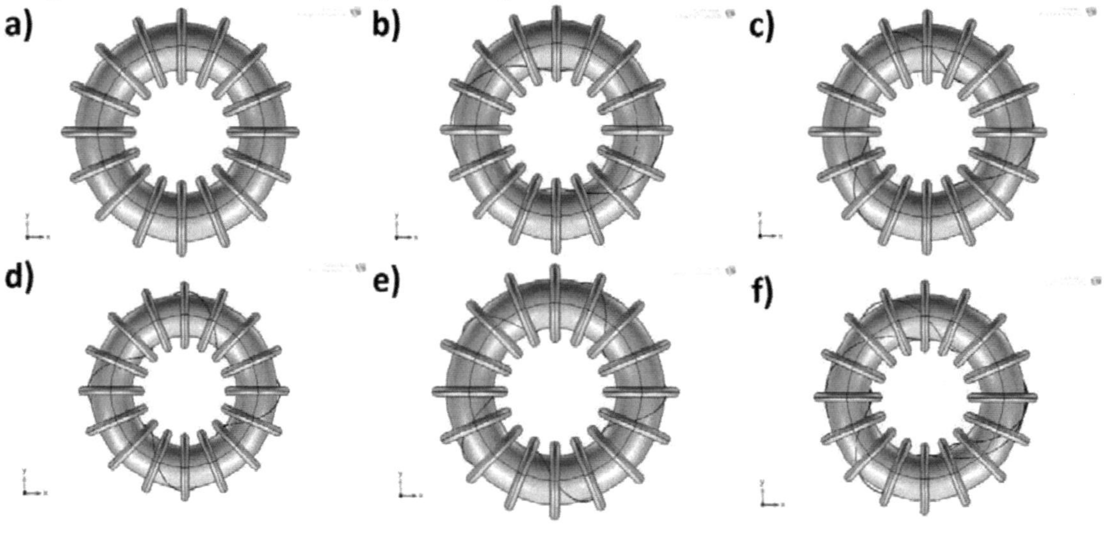


**Figure 2.** Resonance helical perturbation Coil design as seen from the top of the tokamak. The narrow which wrapped the chamber for pictures b–f is RHF coil; a) without RHF coil; b) RHF coil with $q = 2$; c) RHF coil with $q = 3$; d) RHF coil with $q = 4$; e) RHF with $q = 5$; f) RHF with $q = 2\&3$.

Results of calculation of the ripple for three ratios of the RHF coil current/toroidal coil current 0.1, 1, 10 respectively are shown in figures 3, 4, and 5.

Figure 3 shows the intensity of magnetic field without and with RHF at different modes for the RHF coil current / toroidal coil current = 0.1

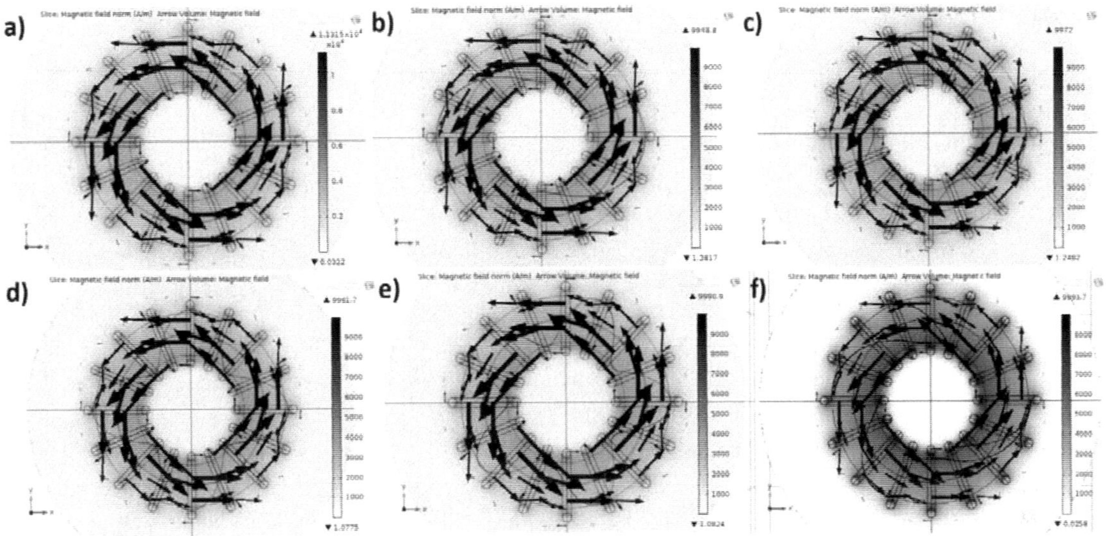


**Figure 3.** The color plot with magnetic field arrows from above tokamak: a) without RHF; b) RHF with $q = 2$; c) RHF with $q = 3$; d) RHF with $q = 4$; e) RHF with $q = 5$; f) RHF with $q = 2\&3$. The arrows and the color intensity show the direction and amplitude of magnetic field entire the tokamak chamber.

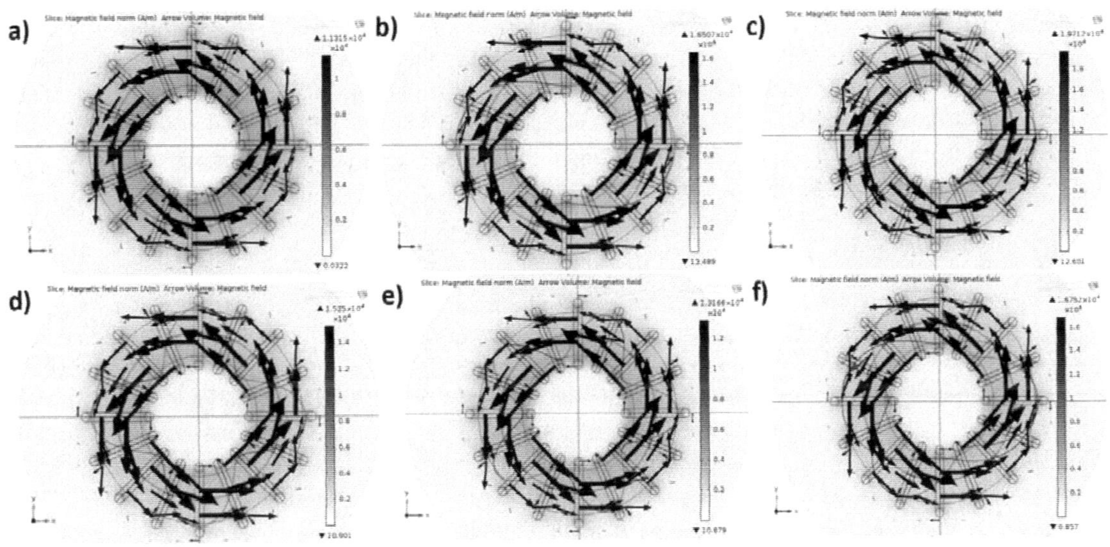


**Figure 4** The color plot with magnetic field arrows from above. tokamak a)without RHF b) RHF with q=2 c) RHF with q=3 d) RHF with q=4 e) RHF with q=5 f) RHF with q=2&3. The arrows and the color intensity shows the direction and amplitude of magnetic field entire the tokamak chamber.

Figure 4 shows the intensity of magnetic field without and with RHF at different q modes for the RHF coil current / toroidal coil current = 1.

Figure 5 shows the intensity of magnetic field without and with RHF at different q modes for the RHF coil current / toroidal coil current = 10.

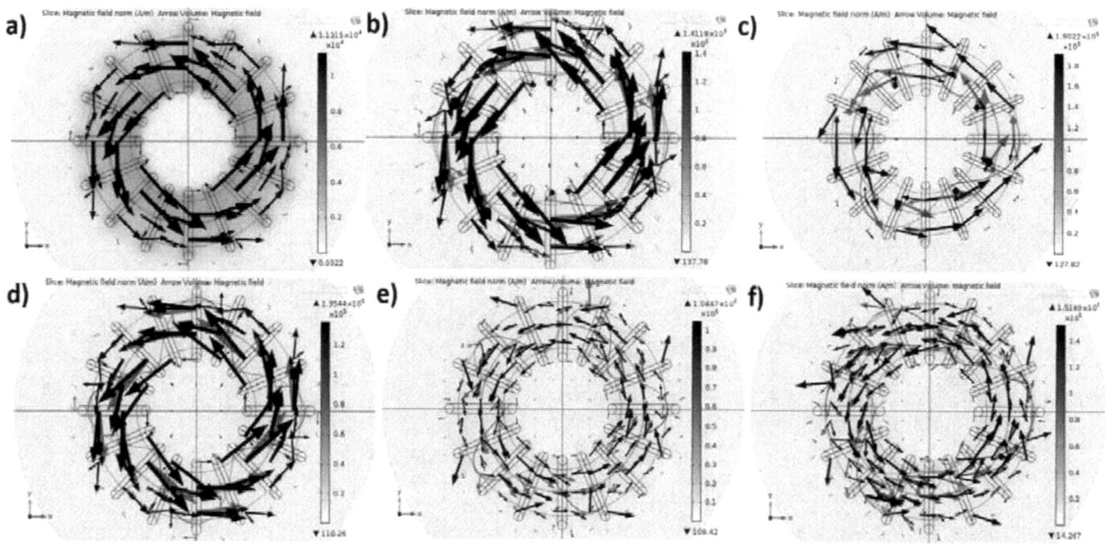


**Figure 5.** The color plot with magnetic field arrows from above tokamak a) without RHF; b) RHF with $q = 2$; c) RHF with $q = 3$; d) RHF with $q = 4$; e) RHF with $q = 5$; f) RHF with $q = 2\&3$. The arrows and the color intensity shows the direction and amplitude of magnetic field entire the tokamak chamber.

The Magnetic field intensity plots in figures 6, 7, 8 show that RHF has affection on toroidal field ripple. As it is clear, the RHF effect increases by increasing current in the RHF coil (RHF coil current /toroidal coil current ratio) as expected. The modulate magnetic field without RHF is because of toroidal magnetic ripple itself. As mentioned before, the magnetic field is stronger within the plane of the toroidal coils than between the coils. This difference of intensity makes magnetic field modulation. This modulation is added to the plot for a better understanding of RHF effects in different $q$ modes.

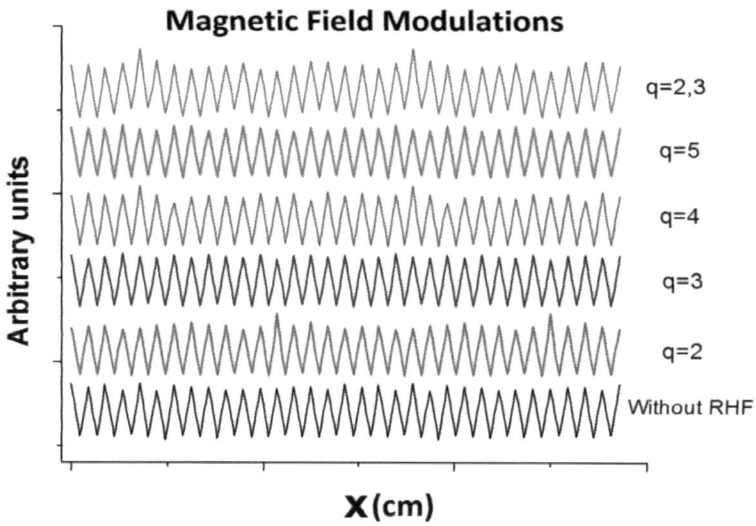


**Figure 6.** The toroidal ripple for different $q$ modes at (RHF coil current / toroidal coil current = 0.1)

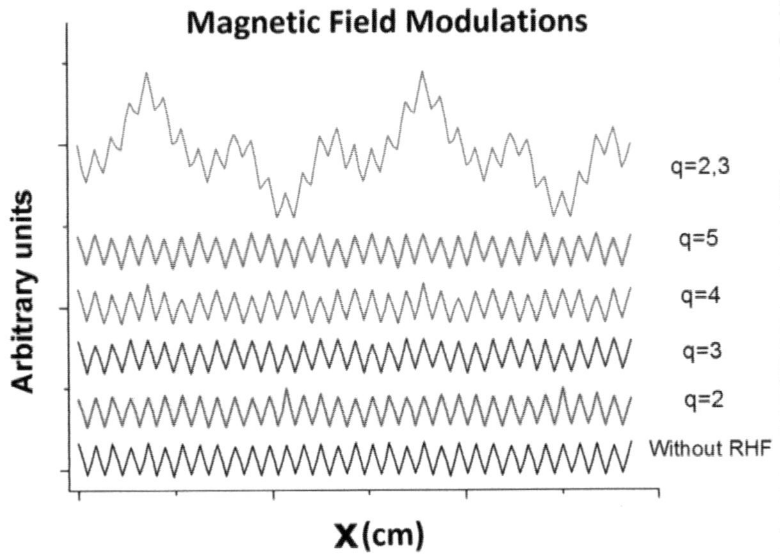


**Figure 7.** The toroidal ripple for different q modes at (RHF coil current / toroidal coil current= 1).

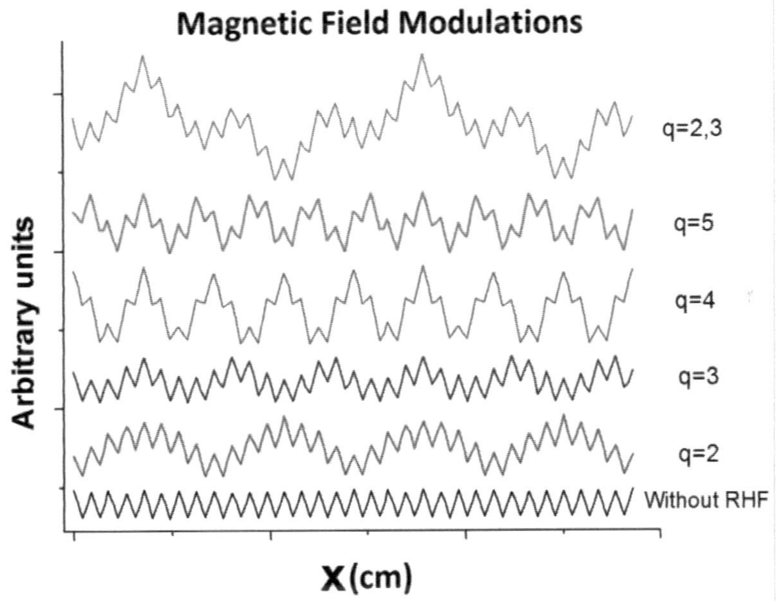


**Figure 8.** The toroidal ripple for different q modes at (RHF coil current / toroidal coil current= 10).

**Table1**. Toroidal ripple error in different RHF currents and modes (%)

| $I_{RHF}$ / $I_{Toroidal\ Coil}$ | Without RHF | Mode2 | Mode3 | Mode4 | Mode5 | Mode2&3 |
|---|---|---|---|---|---|---|
| 0/1 | 1.53 | 1.53 | 1.52 | 1.52 | 1.51 | 1.53 |
| 1 | 1.53 | 1.51 | 1.54 | 1.5 | 1.53 | 1.49 |
| 10 | 1.53 | 1.3 | 1.68 | 1.27 | 1.76 | 1.49 |

IOP Conf. Series: Journal of Physics: Conf. Series **982** (2018) 012004 doi:10.1088/1742-6596/982/1/012004

## 4. Conclusion

The numerical results which obtained from the simulations show that for *q* modes of even numbers, the toroidal field ripple is reduced, and for *q* modes of odd numbers, the toroidal field ripple is increased. Also, we have low discrepancy for mode $q = 2\&3$ (Table 1). However, the modulation of magnetic field lines in high current ratio has high amplitudes which mess up the magnetic field distribution (figure 8).

## References

[1] J. P. Freidberg 2002 *Plasma Physics and Fusion Energy*, MIT Press, Cambridge
[2] J. Narl Davidson 1976 Effect of toroidal field ripple on particle and energy transport in a tokamak, *Journal of Nuclear Fusion* **16** (5) 732–741
[3] M. S. T. Araijo, A. Vannucci and I. L. Caldas 1996 Observation of disruptions in tokamak plasma under the influence of resonant helical magnetic fields *Il Nuovo Cimento D,* **18** (7) 807–821
[4] A. Salar Elahi, M. Ghoranneviss 2009 Measurement of Plasma Energy Confinement Time in Presence of Resonant Helical Field in IR-T1 Tokamak *Journal of Fusion Energy* **28** 394–397
[5] A. Salar Elahi, M. Ghoranneviss 2009 Effects of Resonant Helical Field (RHF) on Equilibrium,Properties of IR-T1 Tokamak Plasma *Journal of Fusion Energy* **28** 416–419
[6] A. Salar Elahi, M. Ghoranneviss 2009 RHF Effect on Shafranov Parameter and Shafranov Shift in IR-T1 Tokamak. *Journal of Fusion Energy* **28** 412–415
[7] B. Mahdavipour, A. Salar Elahi, M. Ghoranneviss 2016 Tokamak Coils Materials and Toroidal Field Ripples Calculation Using the Comsol Multiphysics *Journal of Inorganic and Organometallic Polymers and Materials* **26** (2) 439–445

IOP Conf. Series: Journal of Physics: Conf. Series **982** (2018) 012005 doi:10.1088/1742-6596/982/1/012005

# Numerical analysis of the asymptotic two-scale limit of the plasma-wall transition using a one-dimensional two-fluid model

**T Gyergyek** [1,2], **J Kovačič**[2]

[1] University of Ljubljana, Faculty of electrical engineering, Tržaška 25, 1000 Ljubljana, Slovenia

[2] Jožef Stefan Institute, Jamova 39, 1000 Ljubljana, Slovenia

E-mail: tomaz.gyergyek@fe.uni-lj.si

**Abstract.** A one-dimensional two-fluid model is presented and used for numerical analysis of the asymptotic two-scale limit of the plasma-wall transition. Numerical results confirm that when the problem is treated on the pre-sheath scale, the sheath edge is determined by the electric field singularity. When the problem is approached on the sheath scale, electric field at the sheath edge must be larger than zero in order to obtain any solutions of the model equations. In this case the Bohm criterion is determined by two parameters: the electric field and the ion velocity at the sheath edge.

## 1. Introduction

The plasma-wall transition is one of the oldest in plasma physics [1,2]. A widely accepted of this problem can be summarized in the following way. If neutral plasma is bounded at least on one side by a planar wall that absorbs all particles that reach it and is biased negatively with respect to the plasma potential, a transition layer is formed between the "unperturbed" plasma and the wall, when the potential drop takes place. The "unperturbed" plasma is neutral, its potential is zero and there is no electric field present in such plasma. The transition layer can be divided into 2 parts. The first part is called the pre-sheath. In this region the plasma is neutral, but a finite electric field exists in this region, which accelerates positive particles (ions) towards the wall and negative particles (electrons and possibly negative ions) in the opposite direction. When the velocity of the collective ion motion towards the electrode reaches the ion sound velocity, plasma neutrality breaks down and a space-charge dominated region is formed. This region is called the sheath. In many plasmas the extension of the pre-sheath is much larger than the dimension of the sheath. The length scale $L$ of the pre-sheath is usually some characteristic length that governs the binary processes in this region – usually the mean free path for most frequent collisions. The length scale of the sheath on the other hand is by the rule the Debye length $\lambda_D$, which gives the distance at which the potential perturbation introduced in the plasma is shielded. The ratio of this two lengths $\varepsilon = \lambda_D/L$ is sometimes called the neutrality

IOP Conf. Series: Journal of Physics: Conf. Series **982** (2018) 012005 doi:10.1088/1742-6596/982/1/012005

parameters. The larger is this parameter, stronger local violations of plasma neutrality can occur. In the asymptotic 2 scale limit [1], neutrality parameters goes to zero $\varepsilon \to 0$, since $L >> \lambda_D$. In this limit the analysis of the pre-sheath and of the sheath can be completely separated. In the limit $\varepsilon \to 0$ the boundary between the pre-sheath and the sheath is defined very sharply and it is called the sheath edge. When the sheath edge is approached from the pre-sheath side, electric field gradually increases and becomes infinite (has a singularity) at the sheath edge. When the sheath edge is approached from the sheath side, electric field gradually decreases and becomes zero at the sheath edge.

In this work the asymptotic two-scale limit of the plasma-wall transition is studied numerically using a one-dimensional, steady state, two-fluid model. It is assumed that plasma consists only of electrons and just one species of singly charged positive ions. In the next section the model is presented. In section 3 complete solutions of the model for finite $\varepsilon$ are shown first and then the pre-sheath and the sheath region are studied separately by simply inserting $\varepsilon = 0$ into the model equations. In section 4 some conclusions are presented.

## 2. Model

The model is based on a one-dimensional two-fluid model in steady state:

$$\begin{aligned} &\frac{d}{dx}(n_i u_i) = S_i, \quad \frac{d}{dx}(n_e u_e) = S_e, \\ &m_i n_i u_i \frac{du_i}{dx} = n_i e_0 E - \frac{dp_i}{dx} + A_i - m_i u_i S_i, \\ &m_e n_e u_e \frac{du_e}{dx} = -n_e e_0 E - \frac{dp_e}{dx} + A_e - m_e u_e S_e, \\ &\frac{1}{2} u_i \frac{dp_i}{dx} + \frac{3}{2} p_i \frac{du_i}{dx} = M_i + B_i + \frac{1}{2} m_i u_i^2 S_i, \\ &E = -\frac{d\Phi}{dx}, \quad \frac{d^2\Phi}{dx^2} = -\frac{e_0}{\varepsilon_0}(n_i - n_e). \end{aligned} \tag{1}$$

Here $m_i$ is the ion mass, $u_i$ is the ion fluid velocity, $n_i$ is the ion density, $e_0$ is the elementary charge, $E$ is the electric field, $p_i$ is the ion pressure, $S_i$ is the ion source term, $A_i$ is the ion elastic collision term, which gives the momentum which is transferred from ions to electrons per unit time and per unit volume because of their elastic collisions with electrons, $\Phi$ is the potential, $m_e$ is the electron mass, $u_e$ is the electron fluid velocity, $n_e$ is the electron density, $p_e$ is the electron pressure, $S_e$ is the electron source term, $A_e$ is the electron elastic collision term, which gives the momentum which is transferred from electrons to ions per unit time and per unit volume because of their elastic collisions with ions, $M_i$ is the power density that the ions receive through external heating, $B_i$ gives the density of kinetic energy that is transferred from ions to electrons per unit time because of elastic collisions and $\varepsilon_0$ is the permittivity of the free space. Terms $A_i$ and $A_e$ which describe the exchange of momentum between ions and electrons because of elastic collisions are given by:

$$A_i = -m_e n_e f_{ie}(u_i - u_e), \quad A_e = -m_e n_e f_{ie}(u_e - u_i). \tag{2}$$

Here $f_{ie}$ is the frequency of elastic momentum exchange collisions between ions and electrons (coulomb collisions) and $f_{ie}$ is given by [3]:

$$f_{ei} = \frac{\sqrt{2}}{12\pi^{3/2}} \frac{e_0^4 n_i}{\varepsilon_0^2 \sqrt{m_e}\,(kT)^{3/2}} \ln \Lambda. \tag{3}$$

IOP Conf. Series: Journal of Physics: Conf. Series **982** (2018) 012005 doi:10.1088/1742-6596/982/1/012005

Here $k$ is the Boltzmann constant, $T$ is the mean value of the ion temperature $T_i$ and the electron temperature $T_e$:

$$T = \frac{T_i + T_e}{2}, \tag{4}$$

and $\ln\Lambda$ is the coulomb logarithm, given by:

$$\ln\Lambda = \ln\left(\frac{12\pi\left(\varepsilon_0 kT_e\right)^{3/2}}{\sqrt{n_0 e_0^3}}\right). \tag{5}$$

Here $n_0$ is the density of the unperturbed plasma far away from the electrode. The term $B_i$ is given by [3]:

$$B_i = -\frac{1}{2}kn_e 2\frac{m_e}{m_i} f_{ei}\left(T_i - T_e\right). \tag{6}$$

The source terms $S_i$ and $S_e$ give the difference between the number of created and annihilated ions and electrons per unit volume and per unit time. Since in our model there are only electrons and just one species of singly charged positive ions, it is clear that source terms for both particle species must be identical, $S_i = S_e$. The assumed form depends of the main mechanism of ionization. If the new electron ion pairs are mainly created by ionizing collisions between electrons and neutral atoms, it can be assumed that the source term is proportional to the local electron density:

$$S_e = S_i = \frac{n_e(x)}{\tau}. \tag{7}$$

The ionization time $\tau$ takes into account also the annihilation of both particle species by recombination. The external heating term $M_i$ can be assumed to be a given constant.

The closure of the system (1) is made by the following assumptions. First, the electrons are assumed to be isothermal, so the electron pressure gradient is expressed by:

$$p_e(x) = n_e(x)kT_e, \quad \frac{dp_e}{dx} = kT_e\frac{dn_e}{dx}. \tag{8}$$

Second, it is assumed that the ion heat flux is zero and consequently it does not appear in the fifth equation of the system (1). So the ion pressure gradient term is expressed as:

$$p_i(x) = kT_i(x)n_i(x), \quad \frac{dp_i}{dx} = kT_i\frac{dn_i}{dx} + kn_i\frac{dT_i}{dx} = \kappa kT_i\frac{dn_i}{dx}. \tag{9}$$

Following Kuhn et al [4] the polytropic function $\kappa(x)$ has been defined:

$$\kappa(x) = 1 + \frac{n_i(x)}{T_i(x)}\frac{dT_i}{dn_i}. \tag{10}$$

The following variables are introduced:

$$\lambda_D = \sqrt{\frac{\varepsilon_0 k T_e}{n_0 e_0^2}},\ c_0 = \sqrt{\frac{kT_e}{m_i}},\ L = c_0\tau,\ \varepsilon = \frac{\lambda_D}{L},\ N_i = \frac{n_i}{n_0},\ N_e = \frac{n_e}{n_0},$$
$$V_i = \frac{u_i}{c_0},\ V_e = \frac{u_e}{c_0},\ \Theta = \frac{T_i}{T_e},\ \mu = \frac{m_e}{m_i},\ Z = f_{ei}\tau,\ \Psi = \frac{e_0\Phi}{kT_e},\ \xi = \frac{x}{L}, \tag{11}$$
$$X = \frac{x}{\lambda_D},\ P_i = \frac{p_i}{n_0 kT_e},\ G_i = \frac{M_i\tau}{n_0 kT_e},\ \eta = \frac{e_0 L}{kT_e}E,\ \chi = \frac{e_0\lambda_D}{kT_e}E.$$

If the space coordinate $x$ is normalized to the ionization length $L$, and the equations (2) - (9) are taken into account, the system of the model equations reads:

$$\frac{d}{d\xi}(N_iV_i) = N_e,\ \frac{d}{d\xi}(N_eV_e) = N_e,\ P_i = N_i\Theta,$$
$$N_iV_i\frac{dV_i}{d\xi} = \eta N_i - \frac{dP_i}{d\xi} - \mu Z N_e(V_i - V_e) - V_iN_e,$$
$$\mu N_eV_e\frac{dV_e}{d\xi} = -\eta N_e - \frac{dN_e}{d\xi} - \mu Z N_e(V_e - V_i) - \mu V_eN_e, \tag{12}$$
$$\frac{1}{2}V_i\frac{dP_i}{d\xi} + \frac{3}{2}P_i\frac{dV_i}{d\xi} = G_i - \mu Z N_e(\Theta - 1) + \frac{1}{2}V_i^2N_e,$$
$$\eta = -\frac{d\Psi}{d\xi},\ \varepsilon^2\frac{d^2\Psi}{d\xi^2} = N_e - N_i.$$

If on the other hand the space coordinate $x$ is normalized to the Debye length $\lambda_D$, the system of equations becomes:

$$\frac{d}{dX}(N_iV_i) = \varepsilon N_e,\ \frac{d}{dX}(N_eV_e) = \varepsilon N_e,\ P_i = N_i\Theta,$$
$$N_iV_i\frac{dV_i}{dX} = \chi N_i - \frac{dP_i}{dX} - \varepsilon\left[\mu Z N_e(V_i - V_e) + V_iN_e\right],$$
$$\mu N_eV_e\frac{dV_e}{dX} = -\chi N_e - \frac{dN_e}{dX} - \varepsilon\mu\left[Z N_e(V_e - V_i) + V_eN_e\right], \tag{13}$$
$$\frac{1}{2}V_i\frac{dP_i}{dX} + \frac{3}{2}P_i\frac{dV_i}{dX} = \varepsilon\left[G_i - \mu Z N_e(\Theta - 1) + \frac{1}{2}V_i^2N_e\right],$$
$$\chi = -\frac{d\Psi}{dX},\ \frac{d^2\Psi}{dX^2} = N_e - N_i.$$

In the asymptotic two-scale limit $\varepsilon = 0$ is inserted into systems of equations (12) and (13). The system (12) remains almost the same:

$$\begin{aligned}
&\frac{d}{d\xi}(N_iV_i)=N_e,\ \frac{d}{d\xi}(N_eV_e)=N_e,\ P_i=N_i\Theta,\\
&N_iV_i\frac{dV_i}{d\xi}=\eta N_i-\frac{dP_i}{d\xi}-\mu ZN_e(V_i-V_e)-V_iN_e,\\
&\mu N_eV_e\frac{dV_e}{d\xi}=-\eta N_e-\frac{dN_e}{d\xi}-\mu ZN_e(V_e-V_i)-\mu V_eN_e,\\
&\frac{1}{2}V_i\frac{dP_i}{d\xi}+\frac{3}{2}P_i\frac{dV_i}{d\xi}=G_i-\mu ZN_e(\Theta-1)+\frac{1}{2}V_i^2N_e,\\
&\eta=-\frac{d\Psi}{d\xi},\ 0=N_e-N_i.
\end{aligned}\tag{14}$$

The system (13) on the other hand changes considerably:

$$\begin{aligned}
&\frac{d}{dX}(N_iV_i)=0,\ \frac{d}{dX}(N_eV_e)=0,\ P_i=N_i\Theta,\\
&N_iV_i\frac{dV_i}{dX}=\chi N_i-\frac{dP_i}{dX},\ \mu N_eV_e\frac{dV_e}{dX}=-\chi N_e-\frac{dN_e}{dX},\\
&\frac{1}{2}V_i\frac{dP_i}{dX}+\frac{3}{2}P_i\frac{dV_i}{dX}=0,\ \chi=-\frac{d\Psi}{dX},\ \frac{d^2\Psi}{dX^2}=N_e-N_i.
\end{aligned}\tag{15}$$

In the next section some numerical solutions of the systems of equation (12) - (15) are examined.

## 3. Results

The systems of equation (12) - (15) are strongly nonlinear, so only numerical solutions can be found. Each of the systems is a system of 8 ordinary differential equations for 8 unknown functions of $X$ or $\xi$. For a unique solution also 8 boundary conditions – this means values of the unknown functions at $X = \xi = 0$ - must be specified. Unfortunately here some compromises must be made between physical requirements and limitations imposed by mathematical properties of the systems (12) - (15). In the unperturbed plasma the potential is zero and there should be no electric field. This gives the conditions: $\Psi(0) = \chi(0) = \eta(0) = 0$. At the same place the plasma must be neutral with normalized density of ions and electrons, so $N_i(0) = N_e(0) = 1$. Next are the ion and electron velocity. In the unperturbed region of the plasma there should be no directed ion or electron flow, so this would impose zero velocities at $X = \xi = 0$. But it can be shown [5] that a value $V_i = 0$ results in a singularity by division with zero. So $V_i(0) > 0$ must be selected. For the plots shown in Fig. 1 $V_i(0) = V_e(0) = 10^{-7}$ is selected. In this work we assume that ions are born at rest, so $\Theta(0) = P_i(0) = 0$ is selected. It can also be shown [5] by combining the continuity equations and equations of motion for the ions and electrons that the systems (12), (13) and (15) become singular whenever $V_i \le V_{ith}$ or $V_e \ge V_{eth}$. Here $V_{ith}$ and $V_{eth}$ are ion and electron thermal velocity, given by:

$$V_{ith}=\frac{1}{c_0}\sqrt{\frac{\kappa kT_i}{m_i}}=\sqrt{\kappa\Theta},\ V_{eth}=\frac{1}{c_0}\sqrt{\frac{kT_e}{m_e}}=\frac{1}{\sqrt{\mu}}.\tag{16}$$

The system (14) on the other hand becomes singular when ion and electron velocity reach the values:

$$V_i = \sqrt{\frac{1+\kappa\Theta}{1+\mu\left(\frac{\Gamma+V_e(0)}{\Gamma+V_i(0)}\right)^2}},\ V_e = \sqrt{\frac{1+\kappa\Theta}{\mu+\left(\frac{\Gamma+V_i(0)}{\Gamma+V_e(0)}\right)^2}}. \tag{17}$$

Here Γ is the indefinite integral of the source term over $\xi$, given by:

$$\Gamma = \int N_e(\xi) d\xi, \tag{18}$$

while $V_i(0)$ and $V_e(0)$ are the respective integration constants. If the boundary conditions $V_i(0) = V_e(0)$ are selected, both velocities (17) are equal to the ion sound velocity $V_S$, given by:

$$V_S = \frac{1}{c_0}\sqrt{\frac{kT_e + \kappa kT_i}{m_i + m_e}} = \sqrt{\frac{1+\kappa\Theta}{1+\mu}}. \tag{19}$$

Let us now compare the solutions of the systems (12) and (13) for a finite $\varepsilon$ with the same parameters and boundary conditions. The boundary conditions are: $\Psi(0) = \chi(0) = \eta(0) = \Theta(0) = P_i(0) = 0$, $N_i(0) = N_e(0) = 1$ and $V_i(0) = V_e(0) = 10^{-7}$. The other parameters are: $\varepsilon = 10^{-5}$, $Z = G_i = 0$ and $\mu = 1/3670.482$ (deuterium ions). In Fig. 1 the potential, electric field and ion temperature profiles are shown. In the

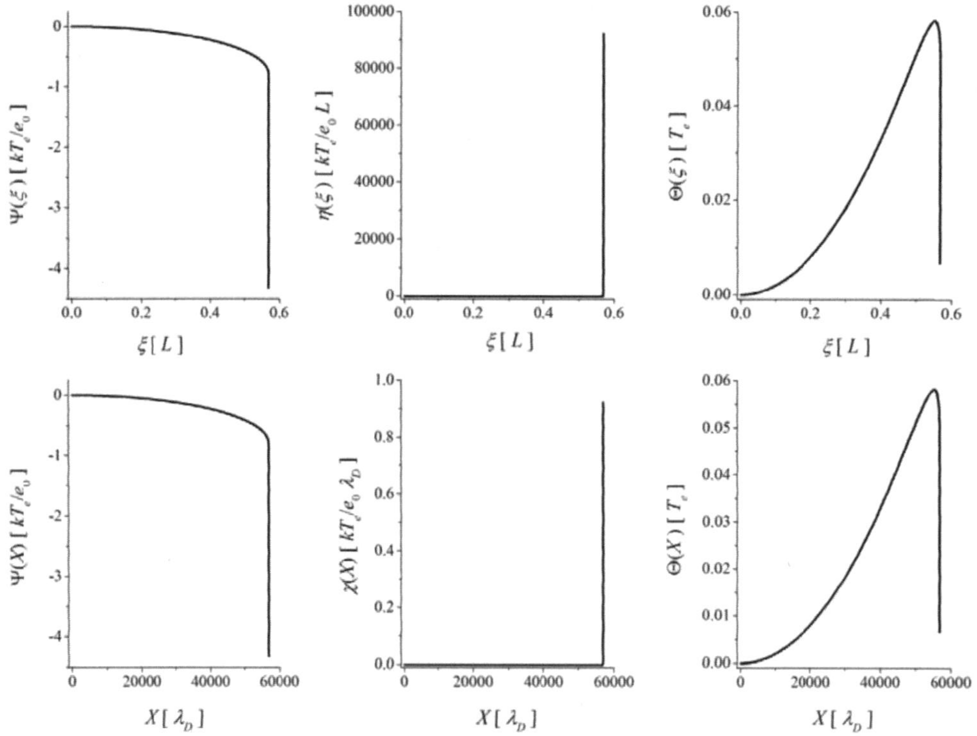


**Figure 1.** In the top plots the profiles of the potential $\Psi(\xi)$, electric field $\eta(\xi)$ and ion temperature $\Theta(\xi)$ obtained from the system (12) are displayed. In the bottom graphs the same profiles found from the system (13) are shown. The boundary conditions and the parameters are: $\Psi(0) = \chi(0) = \eta(0) = \Theta(0) = P_i(0) = 0$, $N_i(0) = N_e(0) = 1$ and $V_i(0) = V_e(0) = 10^{-7}$, $\varepsilon = 10^{-5}$, $Z = G_i = 0$ and $\mu = 1/3670.482$.

plots the profiles of the potential $\Psi(\xi)$, electric field $\eta(\xi)$ and ion temperature $\Theta(\xi)$ obtained from the system (12) are displayed. In the bottom graphs the same profiles found from the system (13) are shown. It can be seen that the solutions of both systems of equations (12) and (13) are identical. The only difference is the scaling of the horizontal scale and of the electric field by factor $\varepsilon$.

Next (Fig. 2) we examine the solutions of the system (14) for the same parameters and boundary conditions, as in Fig. 1, this means: $\Psi(0) = \chi(0) = \Theta(0) = P_i(0) = 0$, $N_i(0) = N_e(0) = 1$, $V_i(0) = V_e(0) = 10^{-7}$, $Z = G_i = 0$ and $\mu = 1/3670.482$. In Fig. 2 all the unknown functions are plotted versus $\xi$. The electric field is shown in the logarithmic scale. It can be seen that close to the sheath edge it increases very sharply for several orders of magnitude. This is in good agreement with the asymptotic two-scale theory [1,2] which claims that in the pre-sheath scale the position of the sheath edge is characterized by the electric field singularity. Ion and electron density $N_i(\xi)$ and $N_e(\xi)$ both decrease monotonically and at the sheath edge they both drop to less than half of the boundary value at $\xi = 0$. Both velocities

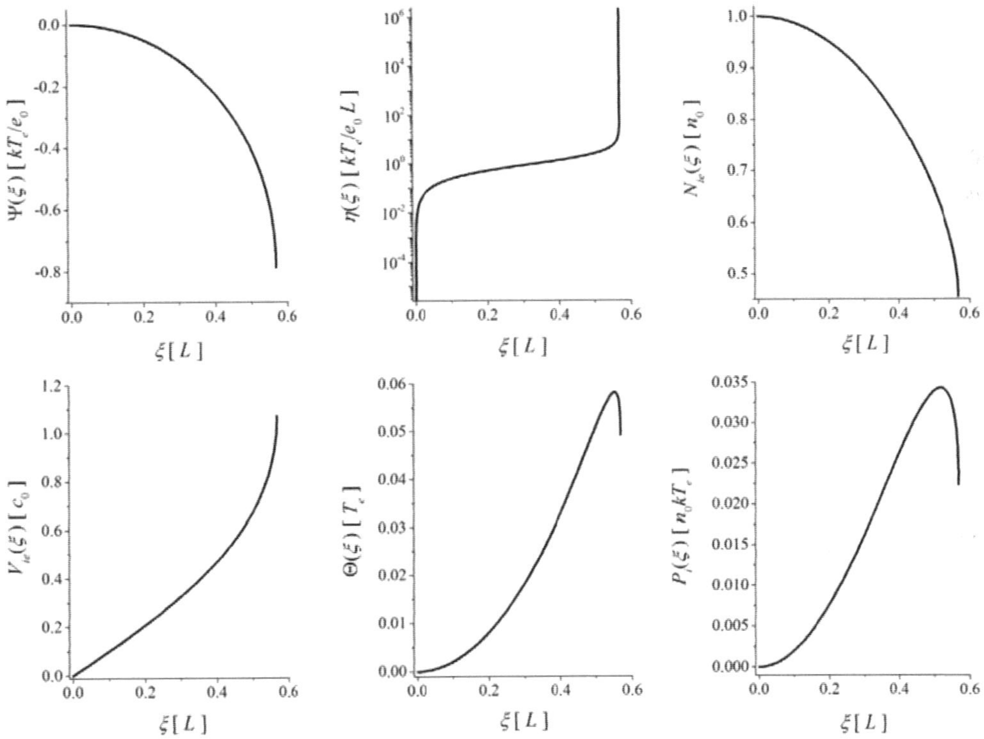


**Figure 2.** Solutions of the system (14) for the following parameters and boundary conditions: $\Psi(0) = \chi(0) = \eta(0) = \Theta(0) = P_i(0) = 0$, $N_i(0) = N_e(0) = 1$, $V_i(0) = V_e(0) = 10^{-7}$, $Z = G_i = 0$ and $\mu = 1/3670.482$.

$V_i(\xi)$ and $V_e(\xi)$ increase monotonically until they reach the ion sound velocity $V_S$, given by (19).

Finally we move to the analysis of the problem on the so-called sheath scale – this means using the system of equations (15). First the issue of boundary conditions arises. At $X = 0$ obviously the sheath edge is located. One might attempt to take the values found from the solutions of the pre-sheath system (14) as boundary conditions for the sheath system (15). The obvious problem with electric field singularity prevents this. Since at the sheath edge the plasma is still neutral the density at the sheath edge can be renormalized to unity and the boundary condition $N_i(0) = N_e(0) = 1$ is taken. Also the potential at the sheath edge can be reset to zero, $\Psi(0) = 0$. The velocities $V_i(0)$ and $V_e(0)$ need some comment. The ion velocity should fulfil the Bohm criterion, so one should take $V_i(0) = V_S$. But in order to do this one must first select the ion temperature $\Theta(0)$ and then find also the polytropic function $\kappa$ using (10). This can be done only after the system (15) has already been solved. Since the ion flow at the sheath edge can be expected to be adiabatic, the value $\kappa = 3$ can be selected and then verified from the solution of the system (15) a posteriori using (10). The selection of the electron velocity $V_e(0)$ is almost arbitrary with 2 remarks. First it should not be larger than $V_{eth}$ (see (16)) because in this case the system (15) has a singularity. Second, if $V_e(0) = 0$ is selected, $V_e(X)$ will be zero for all $X$. In this case the singularity (16) is never reached and the point where the integration of the system (15) is stopped must be selected arbitrarily. In order to avoid this, $V_e(0) = 10^{-7}$ is selected. The parameters from previous figures are not changed, so $Z = G_i = 0$ and $\mu = 1/3670.482$ are taken.

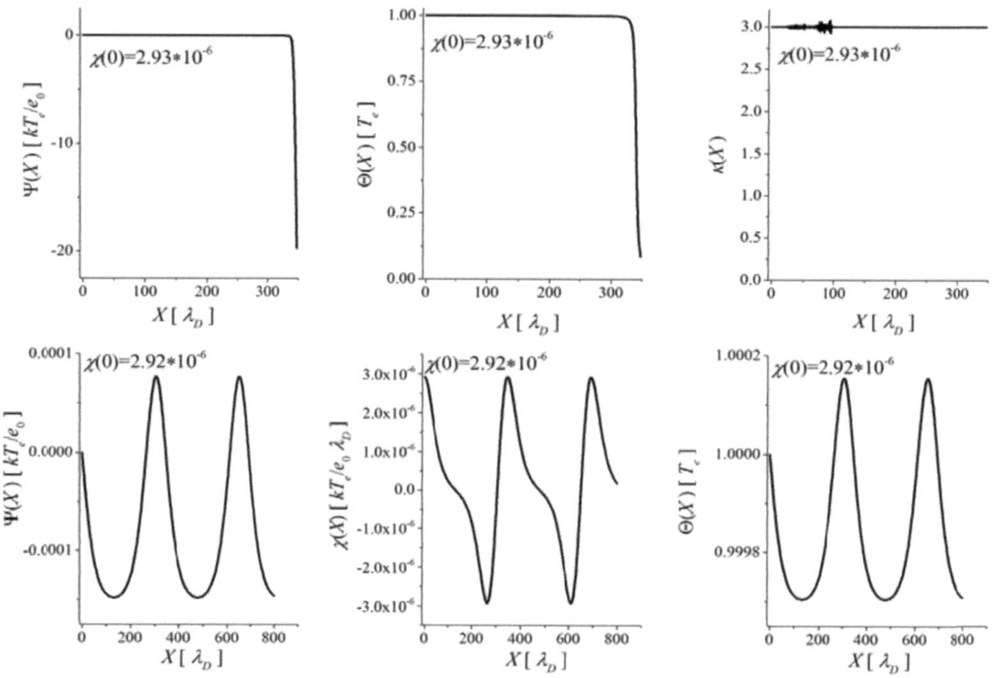


**Figure 3.** In the top plots the potential $\Psi(X)$, the ion temperature $\Theta(X)$ and the polytropic function $\kappa(X)$ are shown for $\chi(0) = 2.93\times10^{-6}$. In the bottom graphs the potential $\Psi(X)$, electric field $\chi(X)$ and ion temperature $\Theta(X)$ are presented for $\chi(0) = 2.92\times10^{-6}$. The other parameters and boundary conditions are: $Z = G_i = 0$ and $\mu = 1/3670.482$, $\Psi(0) = 0$, $V_i(0) = 1.99973$, $V_e(0) = 10^{-7}$, $P_i(0) = 1$, $\Theta(0) = 1$, $N_i(0) = N_e(0) = 1$

The only remaining problem is the electric field $\chi(0)$ at the sheath edge. According to the asymptotic two scale limit [1,2] in the sheath scale the electric field at the sheath edge should be zero. But if $\chi(0)$ is selected exactly 0, the solutions of the system (15) do not move from the boundary values. A small but finite positive value should be selected. So for the top graphs of Fig. 3 the value $\chi(0) = 2.93\times10^{-6}$ is selected. The ion temperature and pressure are $\Theta(0) = 1$, $P_i(0) = 1$ and $\kappa = 3$ is taken. In addition $V_i(0) = 1.99973$ is selected, which corresponds to $V_S$ found from (19). In the top graphs of Fig. 3 the potential $\Psi(X)$, ion temperature $\Theta(X)$ and polytropic function $\kappa(X)$ are presented. It can be seen that the solutions are monotonic and the polytropic function has a constant value 3. The curves shown in the bottom plots of Fig. 3 are also obtained from the system (15) for the same parameters and boundary conditions, only the electric field at the sheath edge is slightly decreased.

The value $\chi(0) = 2.92\times10^{-6}$ is selected. In this case the solutions are oscillatory. This means that at $X = 0$ the Bohm criterion is not fulfilled anymore. For the system (15) the Bohm criterion is determined by two quantities $V_i(0)$ and $\chi(0)$. If one of these boundary conditions is decreased, the other should be increased if one wants to get monotonic solutions.

In Fig. 4 the smallest $V_i(0)$ which results in a monotonic solution of the system (15) is plotted versus $\chi(0)$. The boundary condition $\chi(0)$ is varied over 10 orders of magnitude, so the horizontal scale is logarithmic. The parameters and boundary conditions are: $Z = G_i = \Psi(0) = 0$, $\mu = 1/3670.482$, $V_e(0) = 10^{-7}$ and $N_i(0) = N_e(0) = 1$. Four values of $\Theta(0)$ are selected. The horizontal lines in the plots indicate the values of the respective ion sound velocities, calculated with equation (19) using $\kappa = 3$. It can be seen that for boundary electric fields up to roughly $\chi(0) \sim 2\times10^{-6}$ the boundary ion velocity $V_i(0)$ must slightly exceed the respective ion sound velocity $V_S$ in order to obtain monotonic solutions. For larger $\chi(0)$ the velocity $V_i(0)$ can be slightly smaller than the respective $V_S$.

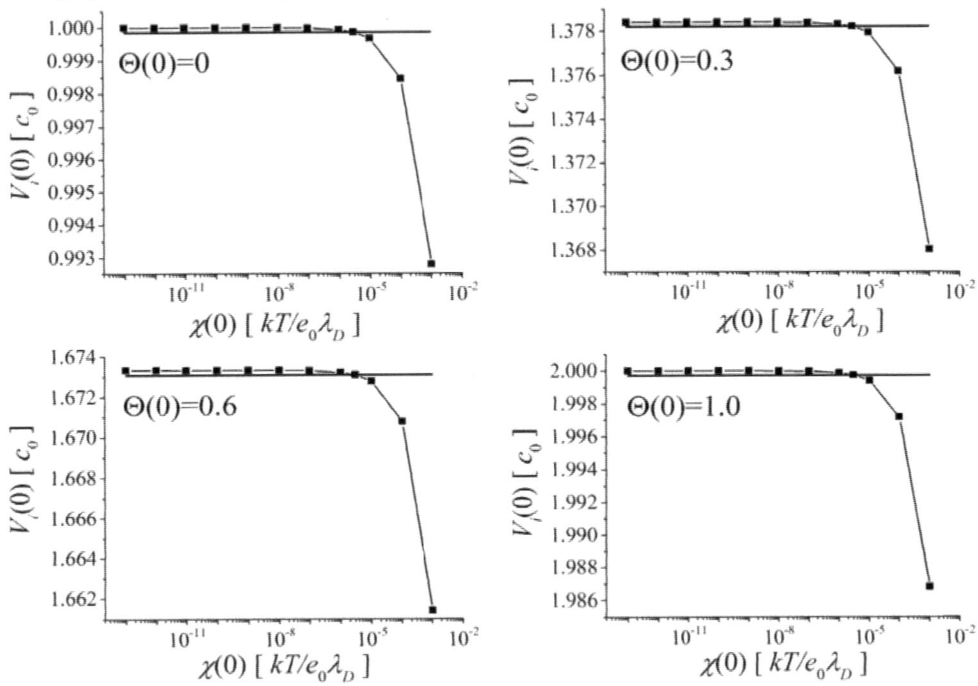


**Figure 4.** The minimum ion velocity $V_i(0)$ which results in a monotonic solution of the system (15) versus $\chi(0)$. The parameters and boundary conditions are: $Z = G_i = \Psi(0) = 0$, $\mu = 1/3670.482$, $V_e(0) = 10^{-7}$ and $N_i(0) = N_e(0) = 1$ and 4 values of $\Theta(0)$ are selected. The horizontal lines indicate the values of the respective $V_S$, calculated with equation (19) using $\kappa = 3$.

## 4. Conclusions

A one-dimensional two-fluid model has been presented and used for numerical analysis of the asymptotic two-scale limit [1,2]. Numerical results confirm that when the problem is treated on the pre-sheath scale, the sheath edge is determined by the electric field singularity. When the problem is approached on the sheath scale, electric field at the sheath edge must be larger than zero in order to obtain any solutions of the model equations. In this case the Bohm criterion is determined by two parameters: the electric field and the ion velocity at the sheath edge. If a very small value of the electric field at the sheath edge is selected, the ion velocity at the sheath edge must be increased slightly above the corresponding ion sound velocity if monotonic solutions are to be obtained. For larger values of the electric field at the sheath edge the ion velocity at the sheath edge can be even slightly smaller than the corresponding ion sound velocity.

**Acknowledgements**
This work has been partially supported by the grant P2-0073 of the Slovenian Research Agency.

## References

[1] Riemann K U 1991 *J.Phys. D: Appl. Phys.* **24** 493
[2] Riemann K U 2000 *J. Tech. Phys. Special Issue* **41** 89
[3] Freidberg J »Plasma Physics and Fusion energy«, Cambridge University Press, Cambridge, New York, (2007).
[4] Kuhn S, Riemann K U, Jelić N, Tskhakaya D D, Tskhakaya D Jr, Stanojević M 2006 *Phys.Plasmas* **13** 013503
[5] Gyergyek T, Kovačič J 2016 *Phys. Plasmas* **23** 063510

IOP Conf. Series: Journal of Physics: Conf. Series **982** (2018) 012006 doi:10.1088/1742-6596/982/1/012006

# Modelling of Argon Cold Atmospheric Plasmas for Biomedical Applications

**M Atanasova[1], E Benova[1], G Degrez[2], and J J A M van der Mullen[2]**

[1] Department for Language Teaching and International Students, Sofia University, 27 Kosta Loulchev St., 1111 Sofia, Bulgaria
[2] Department of Applied Sciences, Université Libre de Bruxelles, CPI 165/04, 50 av. Fr. Roosevelt, 1050 Brussels, Belgium

E-mail: mariana.i.pencheva@gmail.com

**Abstract**. Plasmas for biomedical applications are one of the newest fields of plasma utilization. Especially high is the interest toward plasma usage in medicine. Promising results are achieved in blood coagulation, wound healing, treatment of some forms of cancer, diabetic complications, etc. However, the investigations of the biomedical applications from biological and medical viewpoint are much more advanced than the studies on the dynamics of the plasma. In this work we aim to address some specific challenges in the field of plasma modelling, arising from biomedical applications – what are the plasma reactive species' and electrical fields' spatial distributions as well as their production mechanisms; what are the fluxes and energies of the various components of the plasma delivers to the treated surfaces; what is the gas flow pattern? The focus is on two devices, namely the capacitive coupled plasma jet and the microwave surface wave sustained discharge. The devices are representatives of the so called cold atmospheric plasmas (CAPs). These are discharges characterized by low gas temperature – less than 40°C at the point of application – and non-equilibrium chemistry.

## 1. Introduction

Plasmas for biomedical applications are one of the newest fields of plasma utilization. Particularly plasma use in health care dates from about the beginning of this century. Promising results are achieved already in blood coagulation, wound healing and treatment of several forms of cancer [1]. There are indications that plasma medicine could offer solutions for lesions that cannot be treated otherwise, like diabetic complications and other severe ulcers [2].

However, the investigations of the biomedical applications from biological and medical viewpoint are much more advanced than the studies on the dynamics of the plasma – the configuration effect, i.e. the practically appropriate design and its effect on the plasma properties, the chemical composition, the transport of species, etc. In this work we aim to address some specific challenges in the field of plasma modelling, which arise from biomedical applications. Three main problems stand out in this research area: what are the plasma reactive species' and electrical fields' spatial distributions and what are the production mechanisms; what are the fluxes and energies of the various species that the plasma delivers to the treated surfaces (cells, tissues, etc.); what is the gas flow pattern?

The recent and most widely used, especially for medical applications, devices are the so called cold atmospheric plasmas (CAPs). These are discharges characterized by low gas temperature – less than

IOP Conf. Series: Journal of Physics: Conf. Series **982** (2018) 012006 doi:10.1088/1742-6596/982/1/012006

40°C at the point of application – and non-equilibrium chemistry. Typically they are RF apparatuses [3]. Microwave devices working at higher gas flow (jets) can also be employed [4].

So far, an extensive theoretical and experimental study has been done only on the *plasma needle* [5]. Various other RF atmospheric plasma torches with different design are tested for multiple applications but not well studied and characterized (see for instance [6] and the references therein). In this study we focus on two of these devices, namely the capacitive coupled plasma (CCP) jet shortly denoted by *plasma shower* [7] and on a plasma device, which has barely been investigated for biomedical applications – the surface wave sustained discharge (SWD) [8].

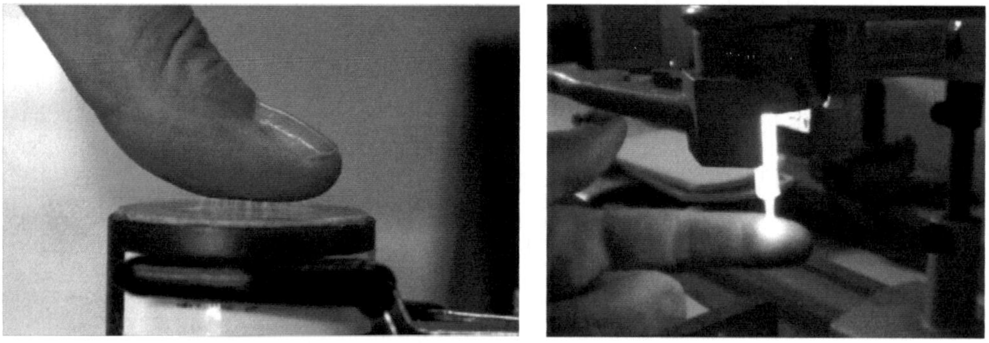

**Figure 1.** The atmospheric pressure capacitively coupled plasma jet with multi-perforated electrodes, shortly denoted by *plasma shower* (left) and a tube-plasma configuration of a surface-wave discharge (right)

Both of the considered devices operate in argon at atmospheric pressure ($p$ = 1 bar) as it is shown in figure 1. However, the difference in the frequency of the power coupling mechanism induces a big difference in plasma properties.

## 2. Modelling Approaches

### *2.1. The SWD Model*

The plasma in a surface wave discharge is created and sustained by an electromagnetic wave. The wave is generated by a microwave source. It is with a frequency of 2.45 GHz. The wave propagates along the interface between the plasma and the dielectric tube (with relative permittivity $\varepsilon_d$) keeping the maximum of the field intensity at the plasma surface; hence the name of the wave. The surface electromagnetic wave creates the plasma and at the same time the plasma is a part of the waveguide structure for the wave propagation. The wave gradually transfers its power to the plasma, more precisely to the electrons. The power decreases with the distance from the launcher and so does the plasma density. The energy the electrons absorb from the wave is expended on different elementary processes which sustain the plasma. When the absorbed power is not enough to compensate the losses associated with radiation, diffusion and recombination the discharge ends.

This type of plasma can be sustained in a glass tube or directly in open air as a torch. Figure 2 shows schematic representation of the discharge and the cross sections of the two geometries, which have axial symmetry.

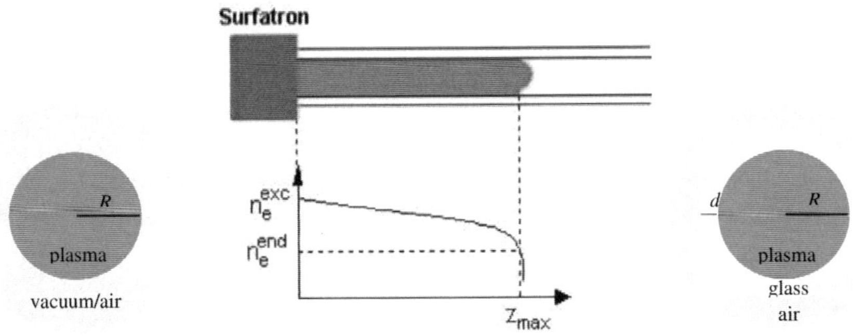


**Figure 2.** Schematic representation of the a discharge created and sustained by electromagnetic surface wave (centre) in two configurations – plasma torch (left) and plasma in tube (right)

A physically sound theoretical model for a discharge created by a surface wave require account in an interconnected manner of both aspects of the problem, namely, the *electromagnetic* and the *plasma* i.e. *kinetic* aspect. The former accounts for the change of wave characteristics – dispersion, spatial distribution of the field components, energy dissipation, while the later takes into account all the processes that ensure the balance between the creation and destruction of particles in the plasma as well as the energy balance of the particles. The full model provides a complete self-consistent description of plasma and wave features. The development of such a model, however, is a rather complex task. Hence usually at some stage of the work it is necessary to adopt some simplifying assumptions. In our model it is assumed that the plasma is in a steady state and radially averaged plasma density is used, building one-dimensional axial model.

### *2.2. The CCP Model*

A model of an RF capacitively coupled discharge with the form of a shower has been developed and used to obtain key plasma parameters as well as to gain insight into the physical processes that determine the plasma behaviour.

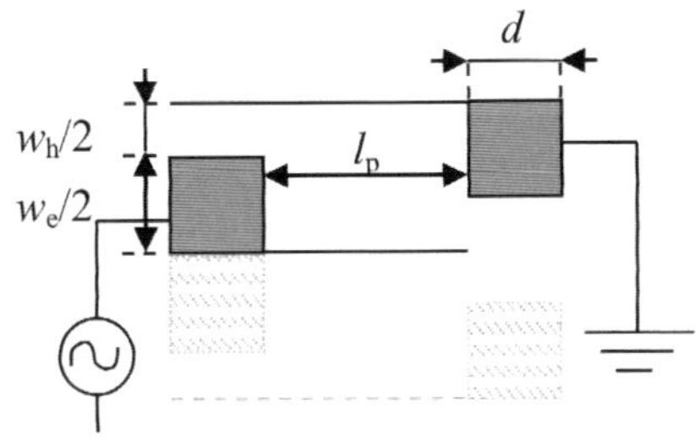


**Figure 3.** Schematic of a unit-cell representing the perforated electrodes.

The configuration we are aiming at should mimic the plasma shower as well as possible. However this shower is intrinsically 3D while our model is 2D. To describe reality in a reasonable way we replaced the shower plates by two parallel plates equipped with slits. These slits are shifted with respect to each other in order to represent the fact that the shower holes are not in line with each other. The width of these slits is the same as the diameter of the holes. The flow rate is selected such that the gas velocity through these slits is the same as that through the shower holes.

In the modelling the device is considered as a periodic 2D Cartesian structure. The structural unit-cell, shown in the figure, includes half of a hole and half of the metal surface between two holes from the powered electrode on the left, the plasma bulk in the middle and a similar electrode pattern but displaced in vertical direction for the grounded electrode on the right. On the top and the bottom of the unit-cell mirror boundary conditions are applied. In the figure, $l_p$ = 2 mm, denotes the distance between the electrodes, $w_h/2 = 0.5$ mm is the half width of the hole, $w_e/2 = 1.0$ mm is the half width of the electrode between two holes, $d = 1$ mm is the depth of the holes.

Three different sub-studies were carried out in order to explore step-by-step the complex cooperation between the different mechanisms: A) a flowless classical parallel plates' geometry, B) a flowless perforated parallel plates' configuration and C) the plasma shower. With this three-stage approach we were able to investigate the role of the sheaths and space charge, the influence of the shower holes on the field distribution and the effect of the flow on the distribution of species [4].

### *2.3. Modelling approaches comparison*

We shell start with a comparison of the main features of the shower and the SWD and a discussion how that affects the corresponding modelling approaches.

Capacitively coupled plasmas (CCP), like the plasma shower, work on electrostatic principles; the functioning of a surface-wave discharge, on the other hand, is based on electrodynamics, i.e. the propagation of EM waves. Thus for a CCP Poisson's law is essential, while the description of the electromagnetic behaviour of SWD is based on the Maxwell equation.

The shower is an electrode-driven discharge; the surface-wave plasma on the other hand is electrodeless. In the RF discharges the sheath next to the electrodes plays an important role in accelerating ions and creating secondary electrons upon collisions with the cathode. This implies that space charges play an important role in the shower plasma, while the surface-wave plasma can be treated as being quasi neutral. The existence of electrodes in the plasma shower implies that surface processes such as secondary electron emission and/or sputtering have to be taken into account. This is not needed for the surface-wave plasma.

In a CCP the electron density and the sheath voltage follow the oscillations of the externally applied electric field and change both in time and space. In a SWD the wave frequency $\omega$ is smaller than the electron plasma frequency $\omega_p$ in the bulk plasma, $\omega < \omega_p$. Since the time scale for the sheath formation is $\omega_p^{-1}$, it follows that the sheath in a SWD remains largely unchanged during one time period of the oscillating EM field. For modelling this means that in a SWD one can consider the electron density as being constant in time [3] so that a stationary model can be used. Moreover we can work in a SWD with space averaged plasma quantities. On the other hand the CCP requires a time-dependent model and for the plasma shower we need at least a two-dimensional geometry to get an adequate description of the plasma in and around the shower holes.

At the industrial driving frequencies of the SWD (typically 2.45 GHz) and the required electron densities (typically $10^{20}$ $m^{-3}$) the wavelengths and skin-depths are comparable or shorter than the typical equipment size. This is not so in the classical CCP that is usually operated at 13.56 MHz. Thus, a uniform amplitude approximation of the oscillating electromagnetic (EM) field of the SWD becomes impossible while in RF discharge it is practically the case.

In the shower the convective flow is essential to bring the active species created in between the plate to the region downstream. In the SWD modeling the gas flow rate can be neglected in many cases.

In both model structures we follow the subdivision of plasma features into three main blocks: configuration, transport and gas discharge chemistry. *Configuration* deals with the impact of the environment on the plasma and thus, among others, with the shape and sizes of the plasma, the boundary conditions and energy coupling modus. *Transport* describes the transport of species, momentum and energy in the plasma (and afterglow). *Chemistry* deals with the creation and destruction of plasma species; hence it has an important impact on the sources and the coefficients that drive and facilitate transport.

A schematic representation, giving the similarities and the differences between the model blocks is given in table 1.

**Table1.** Schematic representation of the plasma features subdivision giving the similarities and differences between the two model approaches

**Abbreviations used**

**EEDF**: Electron Energy Distribution Function, **EK**: Electron kinetics, **HK**: Heavy particle kinetics, **RK**: Radiation Kinetics. The frame in the right lower corner depicts that the collisional-radiative model (CRM) is based on the combination of EK, HK and RK.

| **Configuration** | | **Transport** | | **Chemistry** | |
|---|---|---|---|---|---|
| CCP | SWD | CCP | SWD | CCP | SWD |
| Transient | Stationary | Specific particle balances | Specific particle balances | EEDF | EEDF |
| | | | | EK | EK |
| 2D | 1D | Convection Momentum | no Convection | HK | HK |
| Applied V(t) + space charge | Wave local dispersion relation Wave energy balance | e-Energy balance | e-Energy balance | RK | RK |
| | | | | Collisional-radiative model | |

The numerical approach is also built by means of three main units. They are denoted as chemical kinetics, plasma dynamics and fluid dynamics modules. A schematic representation of the numerical procedure is shown in the figure. The chemical kinetics module, based of Boltzmann solver combined with CRM, provides the necessary data of reaction and transport coefficients. In the case of the plasma shower we work with 4 active species handled by the Boltzmann solver Bolsig+, developed at Toulouse. For the description of the SWD a solver, called *Sofie*, developed in the group of Sofia University is used. It consists of an EEDF solver linked to a CRM.

The plasma dynamics of the shower plasma and SWD is also treated with different software packages. MD2D, a time-dependent two-dimensional fluid model developed at the Technical University of Eindhoven and part of the PLASIMO modelling platform, deals with the configuration–transport aspects of the RF plasma. The SWD dynamics is handled by electrodynamic description of the wave propagation that is self-consistently linked with the kinetic model. In order to account for the effect of the collisions on the wave propagation the full expression for the plasma dielectric permittivity, resulting in complex algebra calculations, is considered.

Regarding the fluid dynamics we calculated the convective velocity entering the convection term in the drift-diffusion-convection equation of MD2D with the Navier-Stokes solver of the PLASIMO platform. As stated before there is no convection modelling applied to the SWD.

## 3. Results and Discussion

### *3.1. The surface-wave plasma*

The investigation of the SWD starts with a study of the propagation characteristics of the surface wave maintaining the plasma [8]. We have analysed the wave dispersion relation for azimuthally symmetric wave at arbitrary ratio of the electron-atom collision frequency and the wave frequency. The analysis includes propagation and damping or attenuation diagrams. The formation of the propagation-damping diagrams for fixed ratio of $\nu_{ea}/\omega$ (corresponding to a given atmospheric pressure plasma column) using particular dispersion curves is demonstrated in figure 5 (a) and (b).

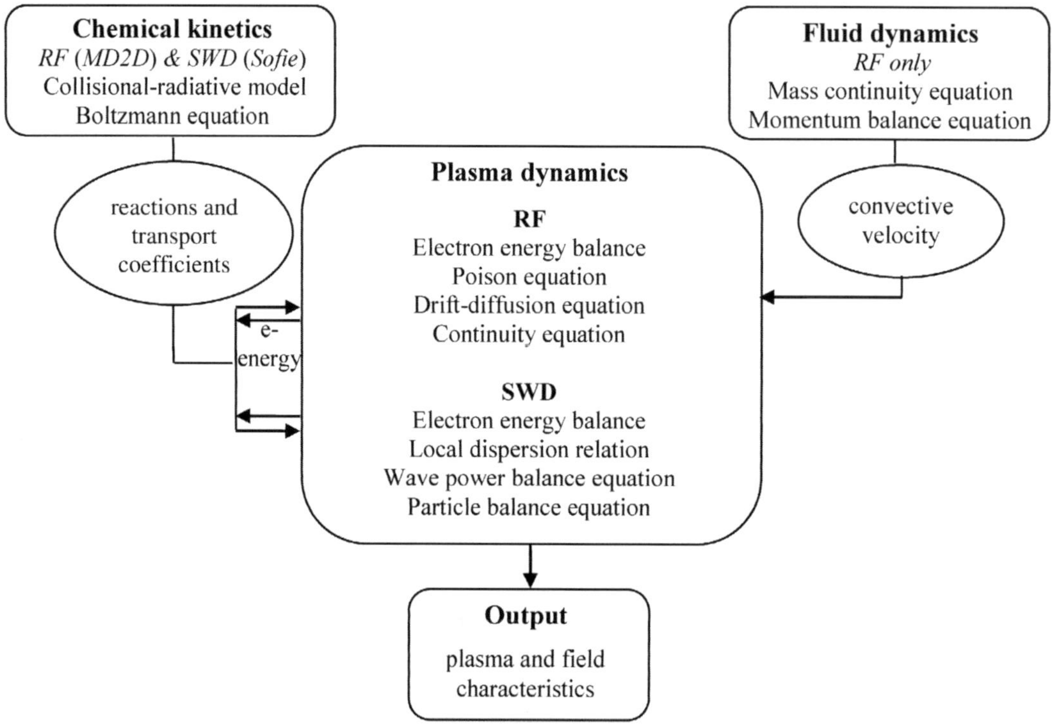


**Figure 4.** Schematic of the numerical calculations showing the communication between the different structure units

Recall that the dispersion relation is normally for uniform plasma and gives the wavelength (via the real part of the wave number, $k'$) as a function of the (variable) wave frequency $\omega$. A more descriptive way to present the dependence is in the normalized form $\omega/\omega_p$ versus $k'R$, the so called *dispersion* diagram (figure 5 (a)). This is equivalent to $\omega$ versus $k'$ since both $\omega_p$ and $R$ are constant. Complementary to it is the *attenuation* diagram giving $\omega/\omega_p$ versus $k''R$ (figure 5 (b)), i.e. the imaginary part of the wave number ($k''$) in dependence on the wave frequency.

In the case of an exciter that launches a wave with *constant* frequency in *non-homogeneous* plasma, the dispersion relation (weakly) depends on the axial coordinate via plasma density and thus $\omega_p$. It must be satisfied at each point along the column. So, in contrast to the classical dispersion, the wave propagation characteristics cannot be described by means of a constant $\omega_p$. Instead, the wave excitation frequency is constant (denoted as $\omega^*$ in this case). The relation between the (changing) $\omega_p$ and the corresponding $k'$ can be presented by means of a *phase* (*propagation*) diagram. Just as in the case of the dispersion diagram it will give $\omega/\omega_p$ versus $k'R$. But the variation in $\omega^*/\omega_p$, is now due to a change in $\omega_p$, not in $\omega$.

In each dispersion and attenuation curve (figure 5 (a) and (b)) there is one point corresponding to wave frequency 2.45 GHz. These points are shown by arrows and they belong to the phase (propagation) and damping diagrams of inhomogeneous plasma sustained by the surface wave with frequency 2.45 GHz. Connecting all such points one actually obtains the phase (propagation) diagram – thick black line in figure 5 (a) and damping diagram – thick black line in figure 5 (b). They can be directly obtain by solving the same dispersion equation at constant $\omega^*$ but varying the plasma density, i.e. $\omega_p$.

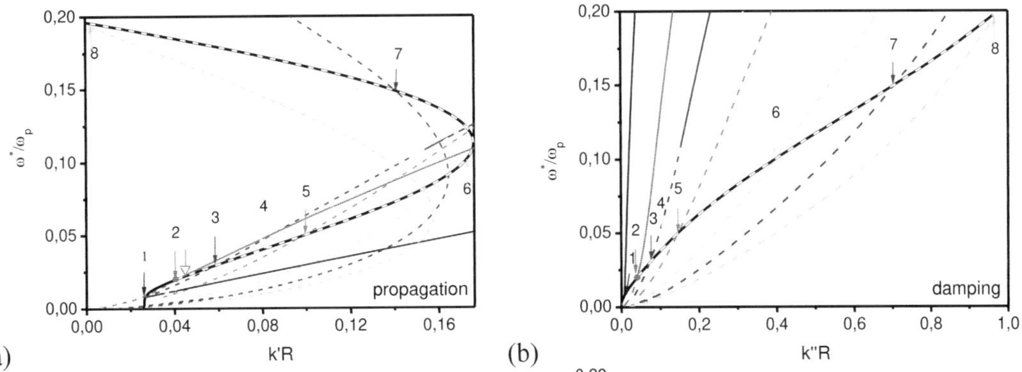


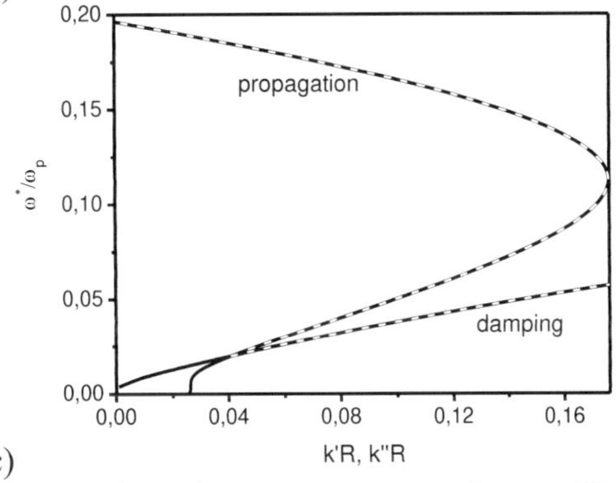


**Figure 5.** Formation of the propagation (phase) (a) and damping (b) diagrams (black dashed lines) for atmospheric pressure plasma column by means of the dispersion and attenuation curves (thin solid and dashed lines). Intersection of the propagation and damping curves (c) at $k' = k''$. The red point in (a) and (b) marks the calculated end of the column corresponding to $k' = k''$. Note the difference in the scales of figures (a) and (b).

Considering that a plasma column can only exist in a region where wave propagation coefficient $k'$ is higher than damping coefficient $k''$ we come to criterion of the plasma column end as the position where $k' = k''$ (see figure 5(c)). The regions presented in figure 5 (c) by solid lines corresponds to plasma existing conditions and all parts with dash-dot lines are just mathematically obtained results for conditions after the end of the plasma. At the plasma column end the electron density $n_e\,(k' = k'')$ is found to be much higher than that corresponding to the turning-back point $n_e$ ( $k'_{max}$, [9]) considered by these authors as a criterion for the column end at the same conditions. The electron density $n_e\,(k' = k'')$ is also much higher than the resonance $n_{e,res}$ ($\omega = \omega_p/\sqrt{1+\varepsilon_d}$), corresponding to the plasma density at the end of low pressure SWD.

The spatial distributions of the components of EM field sustaining a SWD are obtained and the influence of the electron-ion collision frequency $\nu_{ea}$ on the wave components radial distribution is studied. It is found out that the increase of $\nu_{ea}$ facilitates the penetration of the wave into the plasma and diminishes the role of the dielectric tube [7].

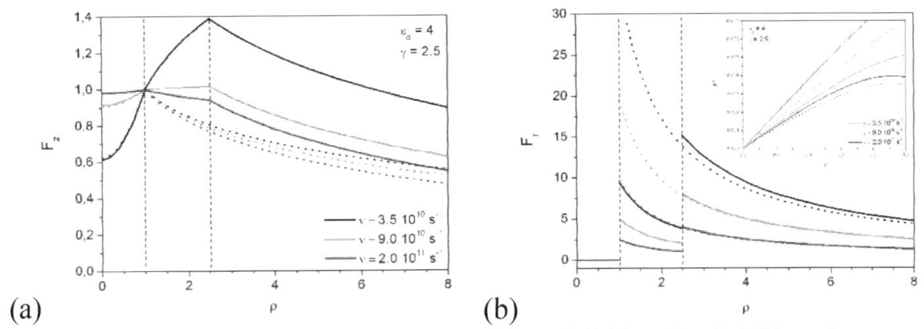


**Figure 6.** Dependence of the radial distribution of the axial (a) and radial (b) surface wave field components on the collision frequency for configuration plasma–dielectric–air – the solid lines and plasma–air – the dotted lines.

IOP Conf. Series: Journal of Physics: Conf. Series **982** (2018) 012006 doi:10.1088/1742-6596/982/1/012006

By means of the complete model the axial distributions of all discharge characteristics can be delivered for the two configurations – plasma inside the dielectric tube and a torch. Figure 7 presents the distributions from the launcher toward the end of the discharge ($z$ = 0 m) along the plasma column of the electron density and temperature, the electron–atom collision frequency and the wave power sustaining the discharge. The red points in figure 7 correspond to the actual end of plasma where the damping coefficient is equal to the propagation coefficient and becomes higher after that points. The gas temperature is considered a fixed parameter and is taken to be $T_g$ = 300 K. The internal radius of the tube (or the plasma radius) equals $R$ = 1 mm, the quartz wall thickness is $d$ = 1 mm and the relative high frequency dielectric permittivity is $\varepsilon_d = 3.8$.

In contrast to low and intermediate pressure, we find for atmospheric pressure SWDs that the wave energy is not always completely dissipated in the plasma (figure 7(c)). The reason is that the plasma sustaining by the wave ends when the damping coefficient becomes higher than the propagation coefficient. At that region the remaining wave power is not enough for sustaining the plasma but the wave can still propagate along the waveguide structure, at least theoretically. The plasma density $n_e$ (figure 7(a)) decreases from the wave launcher to the column end remaining quite high at the end (red points) which is also experimentally observed [10].

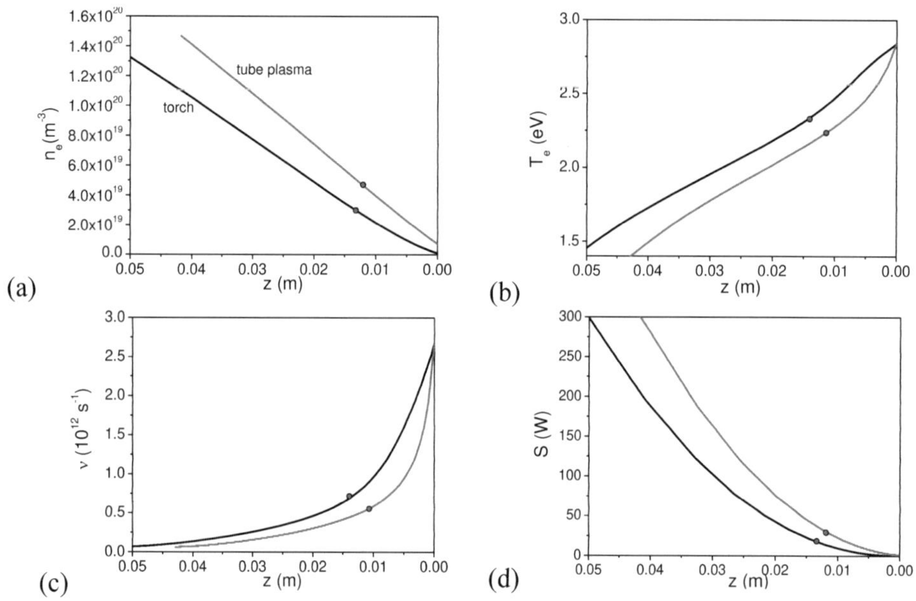


**Figure 7.** Axial distribution of electron density (a), electron temperature (b), electron–atom collision frequency (c) and the wave power sustaining the plasma (d). The red points mark the calculated end of the column corresponding to $k' = k''$.

It is also found that $T_e$ (figure 7 (b)) increases approaching the end of the plasma column. The reason is that towards the plasma end the ionization ratio $\alpha = n_e/N$ is lower and at lower $\alpha$ the tail of the EEDF is more depleted. A depleted tail demands for an enhanced bulk of the EEDF in order to keep the ionization ongoing, which explains the increase of the mean energy of electrons. Similar is the explanation why in the tube plasma where $n_e$ is higher electrons acquire lower $T_e$ and $\nu_{ea}$ (compare figure 7(a) with (b) and (c)).

Finally the axial variation of the atomic states axial distribution is given in figure 8. The predominant species are the molecular ions $Ar_2^+$ which are an order higher than the most populated Ar state Ar(4s) followed by $Ar_3^+$ and $Ar^+$. The density of argon dimmer is also substantial, $n(Ar_2^*) \sim 10^{19}$ $m^{-3}$.

IOP Conf. Series: Journal of Physics: Conf. Series **982** (2018) 012006 doi:10.1088/1742-6596/982/1/012006

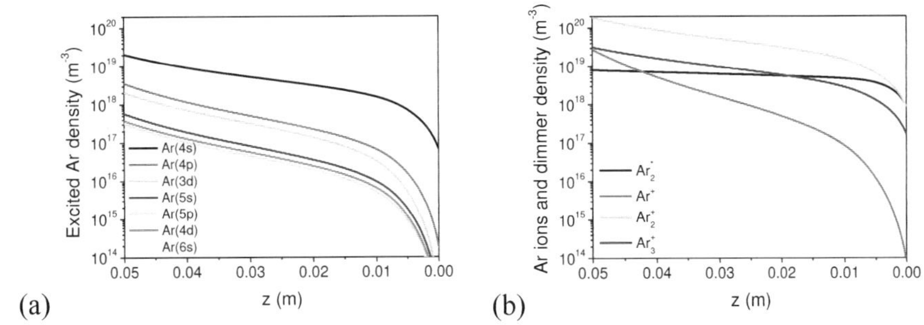


**Figure 8.** Population of Ar excited species (a) and Ar atomic and molecular ions as well as Ar dimmer (b) as function of the axial position *z*.

### 3.2. *The CCP plasma*

The study of the classical parallel plate configuration CCP reveals that strongly non-Maxwellian kinetics is manifested. It results in lowering the direct ionization and a promoting the stepwise ionization. This leads to a two-loop chemistry: 1) the excitation–radiative decay loop and 2) the metastable ionization–recombination circuit. The consequence of this two-loop chemistry and the underlying reaction rates is that the density of the molecular ion is much higher than that of the atomic ion $n(Ar_2^+) >> n(Ar^+)$ and also that the excited species are much more abundant than the charged particles $n(Ar^*) >> n(Ar_2^+)$ as can be seen from figure 9.

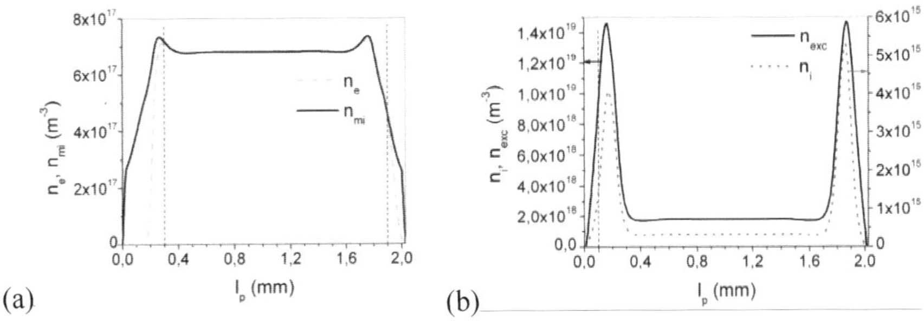


**Figure 9.** Spatial distributions of the densities of (a) electrons and molecular ions, (b) atomic ions and excited atoms at maximum negative potential on the powered electrode

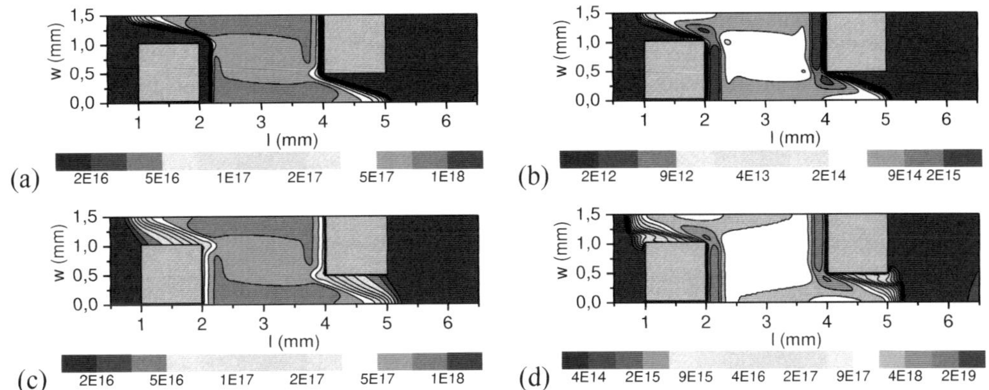


**Figure 10.** Spatial distributions in $m^{-3}$ of the densities of (a) electrons, (b) atomic ions, (c) molecular ions and (d) excited atoms in a plasma shower without flow. All snapshots are taken for maximal negative potential on the powered electrode. Note the difference in the scales.

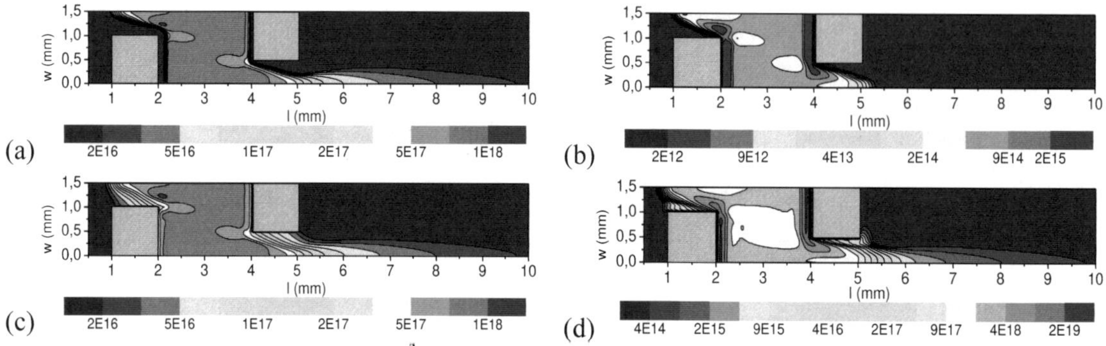


**Figure 11.** Spatial distributions in $m^{-3}$ of the densities of (a) electrons, (b) atomic ions, (c) molecular ions and (d) excited atoms in a plasma shower with flow. All snapshots are taken for maximal negative potential on the powered electrode. Note the difference in the scales.

Finally we find that the flow plays a crucial role: the comparison between figure 10 and figure 11 reveals this. As the most effective outward transport mechanism it is the basis of the formation of the post-discharge region, the region that is of most interest for applications. However as the time scale of the chemical processes is smaller than that of transport even with a flow we get no $Ar^+$ flux in the post-discharge region (see figure 11(b)). In addition to modifying the distribution of the plasma particles in the spatial afterglow the flow recirculation reduces the wall losses and facilitates the ionization and excitation processes in between the plates. This results in higher densities of the plasma constituents there and more homogeneous distribution. Thus it plays a key-role in the stabilization of the discharge.

## 4. Conclusions

Two CAP devices, the *plasma shower* and the *surface-wave discharge* have been investigated in view of their biomedical applications. The main focus was on the three main problems that stand out in this field: the plasma reactive species' and electrical fields' spatial distributions, the fluxes and energies of the various species which the plasma delivers to the treated surfaces and the gas flow pattern. At the same time a discussion on the differences in the modelling approaches was done.

The SWD is found to maintain plasma with higher density $n_e \sim 10^{20}$ $m^{-3}$ while for the CCP it is $n_e \sim 10^{18}$ $m^{-3}$. The predominant species in surface-wave discharge are the charged ones while in the plasma shower they are the excited atoms. Though with lower than the charged particles density the various excited species in the SWD are substantial $n(Ar^*) \sim 10^{17} \div 10^{17}$ $m^{-3}$. The spatial distributions of the plasma constituents and their mean energies for both devices are obtained. The mean energy of the electrons in the SWD is found to be about 1–2 eV versus 3.5 eV in the CCP. The two high frequency excitation fields and the different creating mechanisms related to them are investigated and linked to the different functioning and properties of the discharges. The role of the flow in the CCP is extensively studied and found to play a two-fold role – as the transport mechanism creating the discharge application zone and as a stabilizing factor. The flow in the SWD is not investigated at this stage.

**Acknowledgments**
This work was supported by Bulgarian National Science Fund under Grant No DH 08/8 of 2016.

**References**

[1] Vandamme M *et al.* 2011 *Plasma Med.* **1** 27–43
[2] Fridman G *et al.* 2009 *New J. Phys.* **11** 115012
[3] Laroussi M 2002 *IEEE Trans. Plasma Sci.* **30** 1409–15
[4] Moisan M et al. 2001 International Journal of Pharmaceutics **226** 1–21

[5] Herrmann H W1999 *Phys. Plasmas* **6** 2284–9
[6] Kong M G *et al.* 2009 *New J. Phys.* **11** 115012
[7] Atanasova M *et al.* 2012 *J. Phys. D: Appl. Phys.* **45** 145202
[8] Benova E and Pencheva M 2010 *J. Phys.: Conf. Ser.* **207** 012023
[9] Margot J and Moisan M 1993 *J. Plasma Phys.* **49** 357–74
[10] García MC, Rodero A, Sola A, Gomero A. 2000 *Spectrochim. Acta* **B55** 1733–45

# Bispectral analysis: comparison of two windowing functions

**D Silvagni, C Djerroud, T Réveillé and E Gravier**
Institut Jean Lamour, UMR 7198 CNRS - Université de Lorraine, 54506 Vandoeuvre-lès-Nancy Cedex, France

E-mail: silvagnidavide1@gmail.com, thierry.reveille@univ-lorraine.fr

**Abstract.** Amongst all the normalized forms of bispectrum, the bicoherence is shown to be a very useful diagnostic tool in experimental studies of nonlinear wave interactions in plasma, as it measures the fraction of wave power due to the quadratic wave coupling in a self-excited fluctuation spectrum [1, 2]. In order to avoid spectral leakage, the application of a windowing function is needed during the bicoherence computation. Spectral leakage from statistically dependent components are of crucial importance in the discrimination between coupled and uncoupled modes, as they will introduce in the bicoherence spectrum phase-coupled modes which in reality do not exist. Therefore, the windowing function plays a key role in the bicoherence estimation. In this paper, two windowing methods are compared: the multiplication of the initial signal by the Hanning function and the subtraction of the straight line which links the two extremities of the signal. The influence of these two windowing methods on both the power spectrum and the bicoherence spectrum is showed. Although both methods give precise results, the Hanning function appears to be the more suitable window.

## 1. Introduction

A time series of any fluctuating physical quantity (including fluctuating plasma density and/or potential) may be, to a first approximation, regarded as a superposition of statistically uncorrelated waves and can be described, in part, by its power spectrum which shows the frequency distribution of power of the fluctuations. The power spectrum is of limited value when various spectral components interact with one another due to some nonlinear or parametric process (nonlinear wave-wave interaction in plasma, for example). In such a case, higher order spectral techniques are necessary to accurately and completely characterize the fluctuating signal, since the nonlinearities result in new spectral components being formed which are phase coherent [1]. The detection of such phase coherence may be carried out with the aid of the bicoherence, the normalized form of the bispectrum (third-order spectrum). This is the reason of the actual spreading interest on the bicoherence spectrum in the Fusion research field. In the next Section (II) we are going to briefly present definitions and computational procedures regarding the bicoherence spectrum. In Section III, we show the influence of two windowing functions in both the Fourier and bicoherence spectrum. Lastly, in Section IV we sum up the main results.

## 2. Bispectrum and bicoherence

The bispectrum is the generalization of the power spectrum (second order cumulant spectrum) to the third order. The (discrete) power spectrum $P(k)$, with discrete integer frequency $k$ can be written either as the discrete Fourier transform (DFT) of the second order cumulant $c_2(\tau_1)$, where $\tau_1$ is a discrete lag, or as the expectation of the product of two DFTs $X()$ whose sum frequency is zero [2]:

$$P(k) = DFT(c_2(\tau_1)) = E[X(k)X^*(k)] \tag{1}$$

where $X^*(k) = X(-k)$ denotes the complex conjugate of $X(k)$. In the same way the bispectrum $B(k,l)$, with discrete integer frequencies $k$ and $l$, can be written either as the Double-DFT of the third order cumulant $c_3(\tau_1,\tau_2)$, or as the expectation of the product of three DFTs whose sum frequency is zero:

$$B(k,l) = DDFT(c_3(\tau_1,\tau_2)) = E[X(k)X(l)X^*(k+l)] \tag{2}$$

To clarify how the bispectrum can be used to discriminate between nonlinear coupled waves and spontaneously excited waves we are going to make a simple example. We suppose an experimental situation where only three waves at $\omega_l$, $\omega_k$ and $\omega_m$ satisfy the resonance conditions to have nonlinear wave-wave interaction ($\omega_3 = \omega_1 + \omega_2$ and $k_3 = k_1 + k_2$) [3]. Then the waves have to satisfy the following relation:

$$X(m) = A_{k,l}X(k)X(l) \tag{3}$$

where $m = k + l$ and $A_{k,l}$ is the coupling coefficient. The primary waves at $\omega_k$ and $\omega_l$ interact and generate a third wave at their sum (or difference) frequency $\omega_m = \omega_k + \omega_l$. By writing $X(k) = |X(k)|\exp(i\varphi(k))$, it is clear that the phases must satisfy the condition $\varphi(m) = \varphi(k) + \varphi(l)$ for equation (3) and that the phase of the bispectrum at the frequencies $k$ and $l$ is $\varphi_B(k,l) = \varphi(k) + \varphi(l) - \varphi(m) = 0$ for equation (2) [2]. Therefore, when three spectral components are nonlinearly coupled to each other, the bispectrum phase will not be random at all, although phases of each wave are randomly changing for each realization. Consequently, when a statistical averaging denoted by the expectation operator in (2) is carried out, the bispectrum will take a value different from zero. On the other hand, if the three waves present at $\omega_k$, $\omega_l$ and $\omega_k + \omega_l$ are spontaneously excited independent waves, each wave may be characterized by statistically independent random phases. When the statistical averaging is carried out, the bispectrum will vanish due to the random phase mixing effect [1].

In order to analyze nonlinear wave couplings, the bicoherence is the most appropriate tool. It is defined as:

$$b^2(k,l) = \frac{|B(k,l)|^2}{E[|X(k)X(l)|^2]E[|X(m)|^2]} \tag{4}$$

which is a normalized bispectrum. The bicoherence is often preferred to other normalized bispectrum as it represents the fraction of power at $\omega_m$ due to three-wave coupling and is bounded between 0 and 1 [1].

### *2.1. Numerical computation of the bicoherence*

The bicoherence of deterministic signals can be estimated from a single realisation without averaging only if the noise of the measurement is zero [4]. As this situation is rarely, if ever, encountered in practice, some form of averaging is required to obtain estimates with statistical stability. Averaging can be carried out in one of two ways;

- Multiple realisations of the process can be obtained, and estimates averaged over these realisations (ensemble averaging).
- Only one realisation is used and it is divided up and averaged accordingly.

In the second case (which is the more frequent in practice) we can compute correctly the bicoherence spectrum only if the statistical properties obtained from a single record are the same as those of an ensemble average (ergodic measures) [2].

The computational procedures to compute the bicoherence are as follow.

1) Form $M$ sets of data records of length $N$. They will be denoted as $x^{(i)}(l)$, with $i = 1, \ldots, M$ and $l = 1, \ldots, N$ where $x(l) = x(l\Delta t)$. We assume that each data series $x^{(i)}(l)$ is band-limited and that the sampling interval $\Delta t$ is sufficiently small so that the Nyquist frequency $f_N = 1/(2\Delta t)$ is

larger than any spectral component present in $x^{(i)}(l)$ (if not an antialiasing filter must be applied [5]). Moreover, we assume that the record length $T = N\Delta t$ is large enough to yield sufficient frequency resolution (the elementary bandwidth $\Delta f = 1/T$ must be smaller than the frequency gap between two neighbour spectral components $|f_k - f_{k+1}|$).

2) Subtract the mean value $m = 1/N \sum_{l=1}^{N} x^{(i)}(l)$ from each record $x'^{(i)}(l) = x^{(i)}(l) - m$.
3) Apply an appropriate window to each record $x''^{(i)}(l) = x'^{(i)}(l)\, w(l)$ to reduce spectral leakage [6].
4) Compute the DFT $X^{(i)}(k)$ of each data series:

$$X^{(i)}(k) = \frac{1}{N}\sum_{l=1}^{N} x''^{(i)}(l) \exp\left(-\frac{j2\pi kl}{N}\right) \quad \text{with } k = 1, \ldots, N/2 \text{ and } i = 1, \ldots, M. \tag{5}$$

5) Estimate the bicoherence by:

$$b^2(k,l) = \frac{\left|\frac{1}{M}\sum_{i=1}^{M} X^{(i)}(k)X^{(i)}(l)X^{*(i)}(k+l)\right|^2}{\frac{1}{M}\sum_{i=1}^{M}\left|X^{(i)}(k)X^{(i)}(l)\right|^2 \frac{1}{M}\sum_{i=1}^{M}\left|X^{(i)}(k+l)\right|^2} \tag{6}$$

It is not necessary to compute the bicoherence spectrum over the entire $(k,l)$-plane due to Nyquist frequency limit and due to symmetry relations fulfilled by the bispectrum. It is possible to show that all the bicoherence spectrum is contained in the inner triangular region defined by $0 \le l \le a/2$ and $l \le k \le a - l$ where $a = f_N / \Delta f$ [7].

## 3. Influence of two windowing functions on the bicoherence spectrum

Usually the measured signal is composed of a large number of spectral modes. Therefore, when a nonlinear interaction between two modes (say at $f_k$ and $f_l$) occurs, usually the fraction of power at $f_{k+l}$ due to this nonlinear interaction is not high, since other spontaneously excited modes are present at that frequency. So the bicoherence will usually take small values.

In general, spectral leakage from statistically independent or random phase components will have a similar effect on the bicoherence spectrum as the addition of white random noise [8]. It will lower the fraction of power that is phase-coupled. Likewise, spectral leakage from statistically dependent components will introduce phase-coupled modes which in reality do not exist. If we add these effects to the background noise of the measurement, which also lowers the fraction of power that is phase-coupled, it appears clear that it would be difficult to use the bicoherence spectrum to discriminate between coupled and uncoupled modes. So the windowing function plays a key role in the bicoherence estimation.

Two windowing methods will be analyzed: The Hanning window and the subtraction of the straight line which links the two extremities of the data series (straight line window).

### *3.1 Fourier spectra*

To study the influence of these two window functions on the Fourier spectrum we use a test signal of record length $T = N\Delta t = 2\pi$, $x(l) = \sin(5l\Delta t/4)$ with $l = 1, \ldots, N$, which has $x(1) \neq x(N)$.We already know that a sinus signal with $x(1) = x(N)$ has modes at $k = \pm 1$, so we are looking for the leakage around these modes. The Hanning window consists in the multiplication of the sampled signal by the Hanning function:

$$w_H(l) = \frac{1}{2}\left[1 - cos\left(\frac{2\pi l}{N}\right)\right] = sin^2\left(\frac{\pi l}{N}\right) \quad \text{with } l = 1, \ldots, N \tag{7}$$

The straight line window consists in the subtraction of the straight line which links the two extremities of the sampled signal:

$$w_S(l) = \left[\frac{x(N) - x(1)}{N}\right] l + x(1) \quad \text{with } l = 1, \ldots, N \tag{8}$$

to sampled signal itself.

When these windows are applied to the initial signal, a new windowed signal is obtained with $x(1) = x(N)$, as we can see from figure 1. Then the DFTs of the windowed signals are computed (N = 1024).

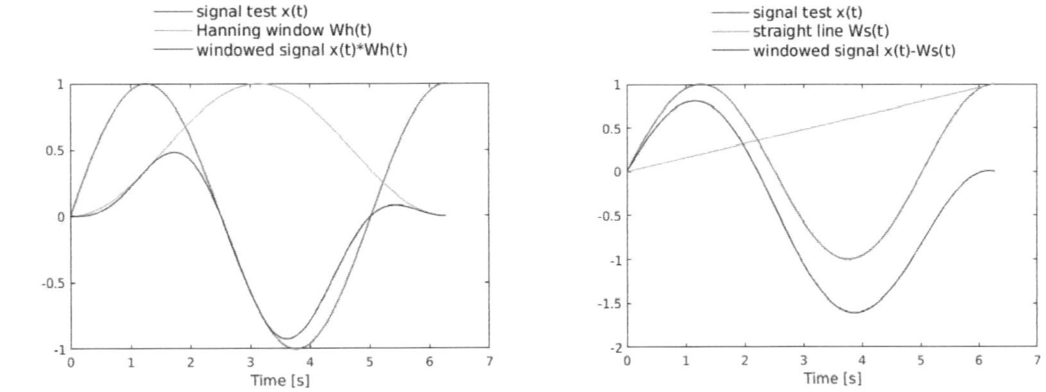


**Figure 1.** Application of the Hanning window (left) and of the straight line window (right) to the signal

As we can see from figure 2 the spectral leakage has a faster decay when the Hanning windowing is applied. Also, we note that the application of the straight line window introduces an additive spectral leakage at $k = 0$ since the mean value of the windowed signal is different from zero. This additive leakage will be removed when, during the bicoherence computation, the subtraction of the mean value of each record is carried out (in this case the windowing operation will be applied before subtracting the mean value).

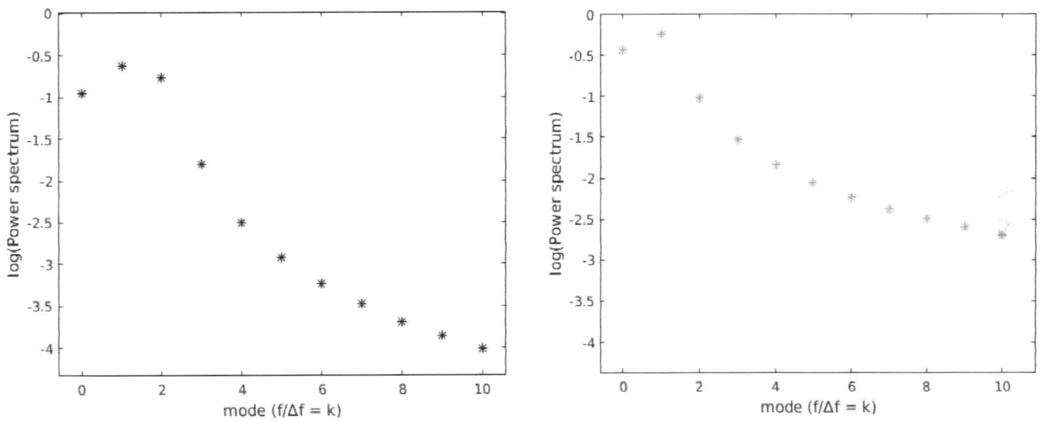


**Figure 2.** Power spectrum of the test signal in logarithmic scale when the Hanning window (left) or the straight line window (right) are applied.

*3.2 Bicoherence spectra*

To study the influence of the two windowing functions on the bicoherence spectrum we generated $M$ = 150 records of such test signal:

$$x(t) = \cos(\omega_b t + \theta_b) + \cos(\omega_c t + \theta_c) + \frac{1}{2}\cos(\omega_d t + \theta_d) + \cos(\omega_b t + \theta_b)\cos(\omega_c t + \theta_c) + \eta(t) \quad (9)$$

where $f_b/f_N = 0.157$, $f_c/f_N = 0.268$, $f_d = f_b + f_c$, $\eta(t)$ is a small amplitude Gaussian noise (–10 dB) and where $\theta_b$, $\theta_c$ and $\theta_d$ are chosen from a set of random numbers which are uniformly distributed over $(-\pi, \pi)$. Each test signal is composed of $N$ = 128 data points. The product term in the test signal will generate the sum and difference frequency waves, which have a phase given by the sum or the difference of $\theta_b$ and $\theta_c$. Thus half of the power at $f_d = f_b + f_c$ is due to the product interaction of waves at $f_b$ and $f_c$ and the other half is independent from the product interaction. The power at $f_a = f_c - f_b$ will be entirely due to the product interaction of waves at $f_b$ and $f_c$ except for a small amount of noise power.

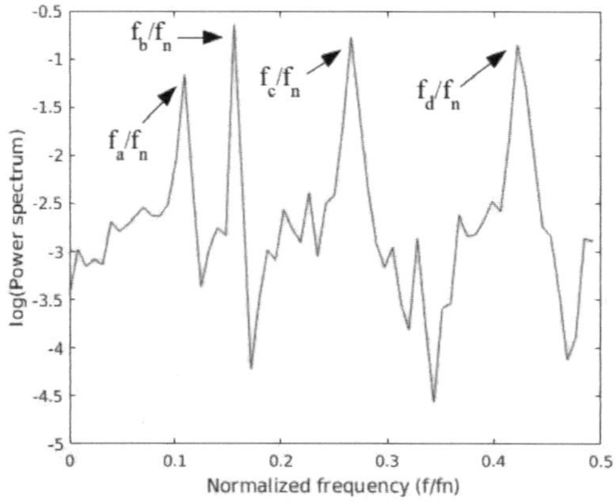


**Figure 3.** Power spectrum in logarithmic scale of the test signal

Figure 3 shows the power spectrum of one record of the test signal. We can clearly see the principal four modes of the signal but we cannot understand whether there are some coupled modes or not (due to the lack of phase information). On the other hand, the bicoherence spectrum allows us to have a clearer picture of wave coupling. Figures 4 and 5 show that the computed bicoherence for the sum interaction was $b^2(b,c) = 0.49$ which implies that only half of the power at $f_d$ is due to the coupling of the waves at $f_b$ and $f_c$. Furthermore the computed bicoherence at the difference interaction was $b^2(-b,c) = b^2(a,c) = 0.97$ which shows that the power at $f_a$ is entirely due to the interaction of waves at $f_b$ and $f_c$. These peaked values are almost the same whether we use the Hanning window or the straight line window. The main difference between the two spectra is due to the effect of spectral leakage when the DFTs are computed. Since the Hanning window reduces markedly the spectral leakage (as we saw from figure 2), the related bicoherence spectrum has a better peak resolution. On the contrary, the straight line window does not reduce sufficiently spectral leakage and so the related bicoherence spectrum has a worst peak resolution, showing peaks with a conic shape which reflects the DFT spectral leakage. This characteristic will affect negatively the bispectral analysis when the signal is composed of many different modes. In this case spectral leakages will make it impossible to clearly distinguish between spontaneously excited waves and nonlinearly coupled waves. Therefore, the Hanning window is found to be more suitable for bispectral analysis than the straight line one.

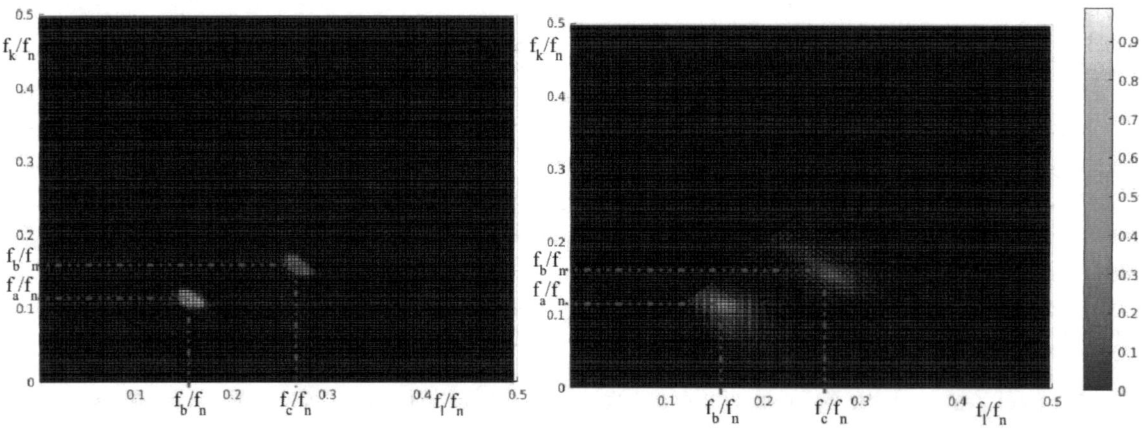

**Figure 4.** 2-D view of the bicoherence spectrum when the Hanning window (left) or the straight line window (right) is applied to a signal which involves the quadratic coupling of two waves.

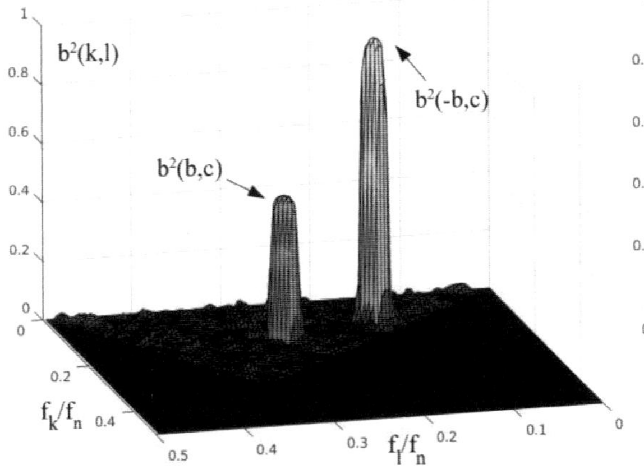


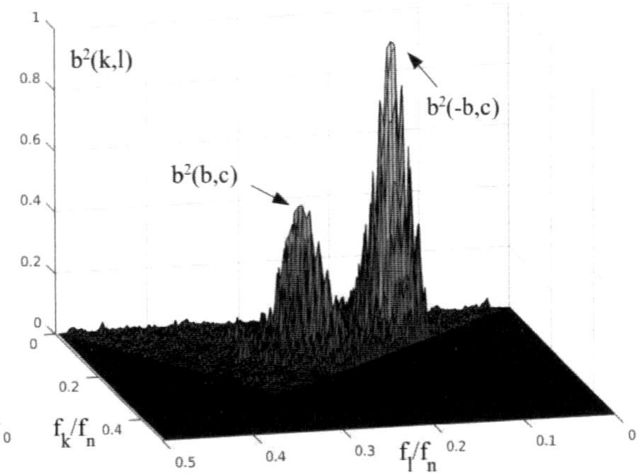


**Figure 5.** 3-D view of the bicoherence spectrum when the Hanning window (left) or the straight line window (right) is applied to a signal which involves the quadratic coupling of two waves.

## 4. Conclusion

To compute the bicoherence spectrum, a window function must be applied to the data series in order to avoid spectral leakage. In general, spectral leakage from statistically independent or random phase components will lower the fraction of power that is phase-coupled [8]. On the other hand, spectral leakage from statistically dependent components will introduce phase-coupled modes which in reality do not exist. If we add these effects to the background noise of the measurement, which also lowers the fraction of phase-coupled power, it is clear that it would be difficult to use the bicoherence spectrum to discriminate between coupled and uncoupled modes. So the windowing function plays a key role in the bicoherence estimation. A comparison between the Hanning window and the so-called straight line window was developed. The analysis of Fourier spectra shows that the application of the Hanning window reduces spectral leakage more than the straight line window. This characteristic is found to have a marked impact on the bicoherence spectrum. When the Hanning window is applied on the test signal, the bicoherence spectrum shows a good peak resolution. On the other hand, when the test signal is windowed by the straight line, the peaks in the bicoherence spectrum have a conic shape, reflecting the spectral leakages in the DFT calculation. Therefore, the straight line window is found not to be suitable for the bicoherence computation.

**Acknowledgments**
The authors wish to thank P. Bertrand and M. Lesur for their support in the development of this paper.

## References

[1] Kim Y C and Powers E J 1979 *IEEE Trans. Plasma Sci.* **PS-7** 120–131
[2] Fackrell J W A 1996 *Bispectral Analysis of Speech Signals* (PhD thesis Edinburgh Research Archive, University of Edinburgh)
[3] Sagdeev R Z and Galeev A A 1969 *Nonlinear Plasma Theory* (New York: Benjamin) Ch. 1
[4] Nikias C and Petropulu P 1993 *Higher Order Spectra Analysis* (New Jersey: PTR Prentice Hall)
[5] Bilinskis I 2007 *Digital Alias-free Signal Processing* (Wiley)
[6] Smith D E, Powers E J and Caldwell G S 1974 *IEEE Trans. Plasma Sci.* **PS-2** 263
[7] Schreier P J and Scharf L L 2006 *Signal Processing* **86** 3321–3333
[8] Chandran V and Stephen L E 1991 *IEEE Trans. on Sign. Processing* **39** 2640

IOP Conf. Series: Journal of Physics: Conf. Series **982** (2018) 012008 doi:10.1088/1742-6596/982/1/012008

# Parametric computational study of sheaths in multicomponent Ar/$O_2$ plasma

**J Hromadka[1], T Ibehej and R Hrach**
Department of Surface and Plasma Science, Faculty of Mathematics and Physics, Charles University in Prague, V Holesovickach 2, 180 00 Prague 8, Czech Republic

E-mail: hromadka.jakub@seznam.cz

**Abstract.** Our study is devoted to sheath structures emerging in Ar/$O_2$ plasma. By means of two dimensional PIC/MCC computer model two configurations were investigated – sheath structure in the vicinity of a cylindrical Langmuir probe for two different biases and changes of the sheath structure when a cylindrical probe passes into a semi-planar probe. It was shown that $O^+$ ions play important role in shielding out negative bias of a solid immersed in Ar/$O_2$ plasma and edge effects of a semi-planar probe on its sheath structure were evaluated.

## 1. Introduction

Plasma that contains more species of both negative and positive ions [1–3], so called multicomponent plasma, is widely used in many industrial applications nowadays, particularly in plasma-assisted technologies (i. e. thin film deposition, plasma etching, etc.). Multicomponent plasma is often a mixture of reactive gases that is chemically active and composed of mutually interacting charged and neutral particles that are out of the thermodynamic equilibrium. Thus, it is very demanding to describe it theoretically and therefore it is often taken advantage of computer modelling approach in multicomponent plasma research [4, 5].

Our contribution is devoted to investigation of sheath structures that are formed in the surroundings of solids immersed in Ar/$O_2$ plasma. The results were obtained by 2D PIC/MCC computer model, which is described in Section 2. Section 3 contains results of the computer simulation of a biased cylindrical Langmuir probe immersed in Ar/$O_2$ plasma. Further, results of a parametric study that captures a transition from a cylindrical to a planar solid substrate are also presented here.

## 2. Modelling technique

The sheath structures in the present contribution were modelled by means of two-dimensional PIC/MCC computer model that was already used to explore sheath structures in purely electronegative plasma [6]. The results of the chemical kinetics model published in [7] were adopted and used as an input for our simulation. The kinetic model takes into account more than 100 reactions between neutral, charged and excited species in Ar/$O_2$ plasma that are described by rate constants and consequently, continuity equations for all species in the ensemble are solved to obtain their number densities. Only the most important charged particles were considered in the PIC simulation: $e^-$, $O^-$, $Ar^+$, $O^+$, $O_2^+$. Neutral particles that were taken into account: Ar, O, $O_2$. Scattering processes were

[1] To whom any correspondence should be addressed.

treated by modified null collision method [8]. The most important interactions between charged and neutral particles that were included in the PIC simulation: [$e^-$, Ar] – elastic scatter, excitation (11.5 eV), ionization (15.8 eV); [$e^-$, O] – elastic scatter, excitation (1.97 eV, 4.19 eV), ionization (13.62 eV); [$e^-$, $O_2$] – elastic scatter, ionization (12.06 eV); [$Ar^+$, Ar] – elastic scatter, charge transfer; [$Ar^+$, O], [$Ar^+$, $O_2$], [$O^-$, O], [$O^+$, O], [$O_2^+$, Ar], [$O_2^+$, $O_2$] – charge transfer; [$O^-$, $O_2$] – elastic scatter. Parameters of investigated Ar/$O_2$ discharge: $E/N$ = 60 Td, pressure 133 Pa, initial Ar to $O_2$ concentration ratio 50:50. The computational domain had dimensions (4.0 x 4.0) · $10^{-2}$ m in all investigated configurations and the computational mesh of 400 x 400 cells was used to solve Poisson's equation.

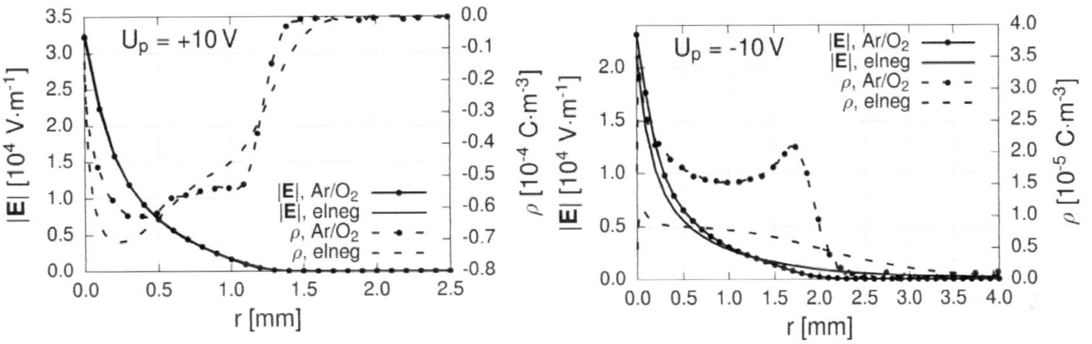


**Figure 1.** Electric field intensity (solid lines) and net charge density (dashed lines) in the vicinity of a single Langmuir probe with $2{\times}10^{-4}$ m radius immersed in Ar/$O_2$ plasma. Two different biases of the probe $U_p$ are presented. The results were compared to those for simplified electronegative plasma [6].

## 3. Computer simulations of Ar/$O_2$ plasma

First of all, the sheath around a single cylindrical Langmuir probe with $2{\times}10^{-4}$ m radius was investigated. Two different biases of the probe with respect to the plasma potential were considered: ± 10 V. The results are depicted in the figure 1 and compared to those for simplified 50 % electronegative plasma [6]. (The term simplified electronegative plasma means in our context argon plasma in which $O^-$ ions are also present; therefore $e^-$, $Ar^+$ and $O^-$ ions represent the only charged particles in this first approximation of Ar/$O_2$ plasma. The amount of $O^-$ ions is given by the electronegativity $\alpha$ which is defined as: $\alpha = n_{O-} / (n_{e-} + n_{O-})$, where $n_{O-}$ denotes number density of $O^-$ ions and $n_{e-}$ that of electrons.) Significant differences are observed for −10 V bias of the probe. In this case $O^+$ ions play important role in shielding out of the negative probe potential since they are lighter than $Ar^+$ ions and more of them get into the sheath. As a result of their presence sheath becomes thinner and electric field increases. On the other hand, the situation in Ar/$O_2$ plasma is quite similar to that in electronegative plasma for +10 V bias of the probe.

The next investigated problem is depicted in the figure 2. The sheath structures during the transition from a cylindrical Langmuir probe of 5 · $10^{-4}$ m radius to a semi-planar probe of $5{\times}10^{-4}$ m thickness were examined for several widths of the semi-planar probe: (2; 6; 11) $\times 10^{-3}$ m. It can be observed that the parameters of the sheath (thickness and magnitude of electric field) near the rounded edges of the semi-planar probe remains the same as in the case of sheath near the cylindrical probe. However, the sheath structure in the middle part of the semi-planar probe significantly changes with

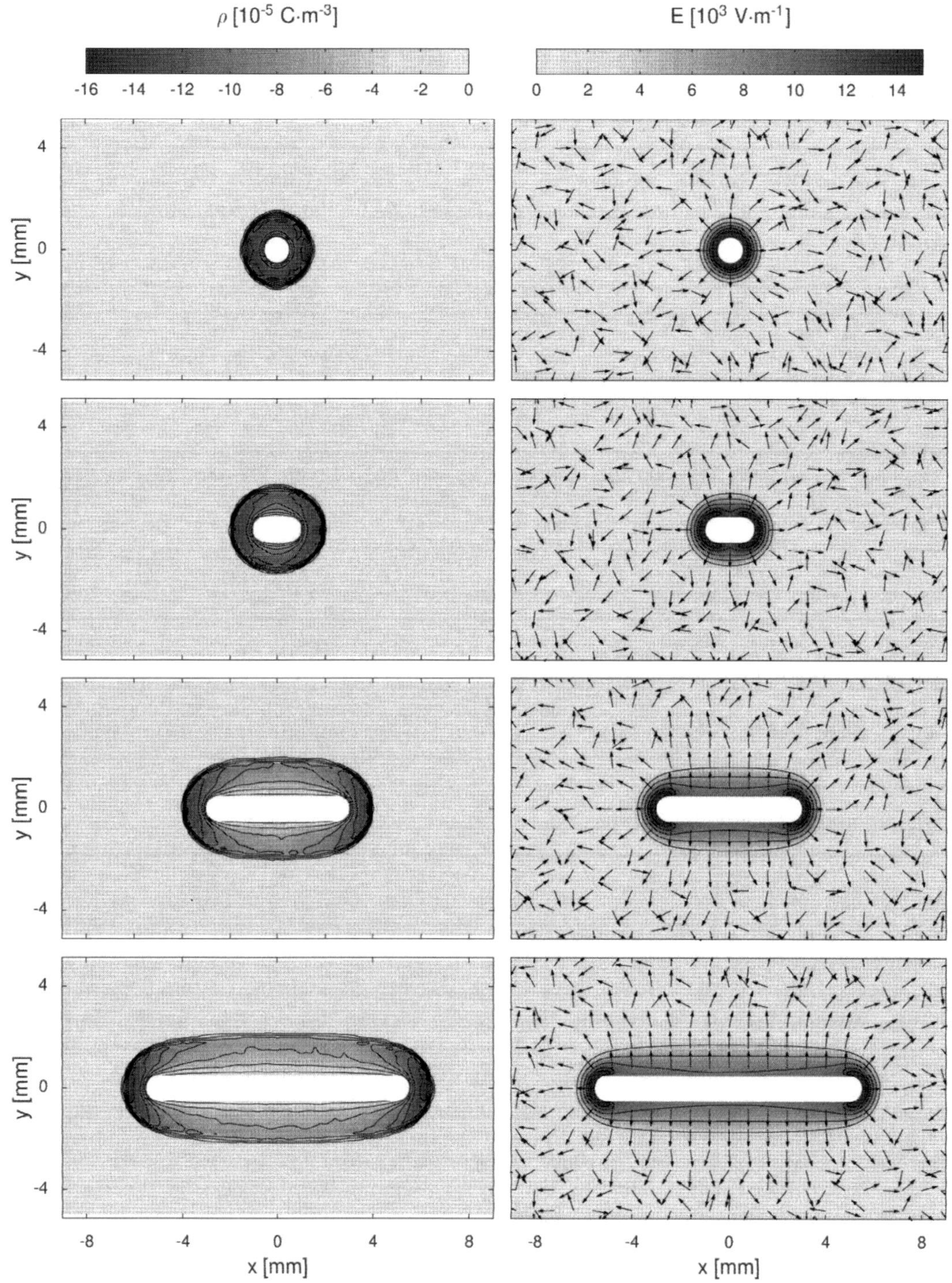


**Figure 2.** Net charge density and electric field intensity in the sheath structure around a cylindrical probe and a semi-planar probe immersed in $Ar/O_2$ plasma. Bias of the probes: +10 V. Only $(1.8 \times 1.0) \times 10^{-2}$ m middle part of the computational domain is depicted. The solid lines mark isolines of the charge density with $1 \times 10^{-5}$ C m$^{-3}$ step and isolines of the electric field intensity with $3 \times 10^{3}$ V m$^{-1}$ step.

increasing width of the probe – electric field intensity decreases and sheath thickness increases. The cylindrical probe sheath solution passes into the solution of a planar probe in this area.

## 4. Conclusions

Two dimensional PIC/MCC computer model of multicomponent $Ar/O_2$ plasma was used to investigate sheath structures in the vicinity of immersed solids. It was proven that $O^+$ ions significantly affect sheaths near the negatively biased solids and it was shown for negatively biased cylindrical Langmuir probe. The edge effects were evaluated for semi-planar probe of different width. It was demonstrated how the sheath structure corresponding to a cylindrical probe passes into the sheath of a planar probe in the middle part of the probe.

## Acknowledgments

The study was supported by the Charles University in Prague, project GA UK No. 220215.

## References

[1] Franklin R N 2001 *J. Phys. D: Appl. Phys.* **34**, 1959
[2] Franklin R N 2002 *J. Phys. D: Appl. Phys.* **35**, 536
[3] Franklin R N 2003 *J. Phys. D: Appl. Phys.* **36**, 34
[4] Birdsall C K, Langdon A B 1991 *Plasma Physics via Computer Modelling* (Bristol: IOP Publishing)
[5] Costin C, Minea T M, Popa G, Gousset G 2010 *J. Vac. Sci. Technol. A* **28**, 322
[6] Hromadka J, Ibehej T, Hrach R, 2015 *Phys. Scr.* **90** 105603
[7] Cerny P, Novak S, Hrach R, Hrachova V 2011 *Eur. Phys. J. Appl. Phys.* **56** 24006
[8] Roucka S, Hrach R 2011 *IEEE Trans. Plasma Sci.* **39** 3244

# Surface-wave-sustained plasma torch for water treatment

**P Marinova[1], E Benova[2], Y Todorova[3], Y Topalova[3], I Yotinov[3], M Atanasova[2] and F Krcma[4]**

[1]*Faculty of Physics, Sofia University, 5 James Bourchier Blvd., 1164,Sofia, Bulgaria*
[2]*DLTIS, Sofia University, KostaLoulchev 27 str., 1111Sofia, Bulgaria*
[3]*Faculty of Biology, Sofia University, 8 DraganTsankov Blvd., 1164Sofia, Bulgaria*
[4]*Brno University of Technology, Purkyňova 118, 612 00 Brno, Czech Republic*

E-mail: plamena_am@abv.bg

**Abstract.** In this study the effects of water treatment by surface-wave-sustained plasma torch at 2.45 GHz are studied. Changes in two directions are obtained: (i) changes of the plasma characteristics during the interaction with the water; (ii) water physical and chemical characteristics modification as a result of the plasma treatment. In addition, deactivation of Gram positive and Gram negative bacteria in suspension are registered. A number of charged and excited particles from the plasma interact with the water. As a result the water chemical and physical characteristics such as the water conductivity, pH, $H_2O_2$ concentration are modified. It is observed that the effect depends on the treatment time, wave power, and volume of the treated liquid. At specific discharge conditions determined by the wave power, gas flow, discharge tube radius, thickness and permittivity, the surface-wave-sustained discharge (SWD) operating at atmospheric pressure in argon is strongly non-equilibrium with electron temperature $T_e$ much higher than the temperature of the heavy particles (gas temperature $T_g$). It has been observed that SWD argon plasma with $T_g$ close to the room temperature is able to produce $H_2O_2$ in the water with high efficiency at short exposure times (less than 60 sec). The $H_2O_2$ decomposition is strongly dependant on the temperature thus the low operating gas temperature is crucial for the $H_2O_2$ production efficiency. After scaling up the device, the observed effects can be applied for the waste water treatment in different facilities. The innovation will be useful especially for the treatment of waters and materials for medical application.

## 1. Introduction

Management of unacceptable bacterial presence and growth in different media, products and water can be ineffective, expensive, time consuming and not always applicable even by the contemporary control methods [1, 2]. Novel methods as plasma treatment of bacteria would be useful in such cases. The low temperature non-equilibrium plasmas have lots of advantages for such applications. Their active components are: electrons, ions, chemically active radicals and excited atoms, electric fields, UV radiation, heating (figure 1).

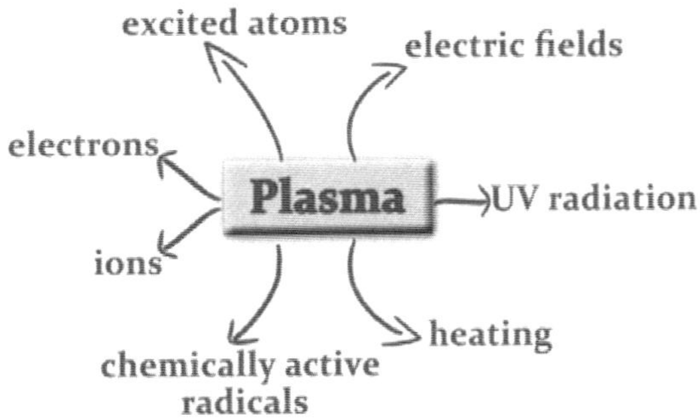


**Figure 1.** Plasma constituents.

The plasma source for these investigations is surface-wave-sustained discharge produced in Argon at atmospheric pressure by surfatron type electromagnetic wave launcher operating at 2.45 GHz (plasma torch). A picture of the plasma torch device used in this pilot study is shown in figure 2.

**Figure 2.** Plasma torch in Argon at atmospheric pressure sustained by surfatron type electromagnetic wave launcher.

This plasma is strongly non-equilibrium: the electron energy distribution function (EEDF) is non-Maxwellian and the temperature of the heavy particles (so called gas temperature $T_g$) is much lower than the electron temperature $T_e$ defined as $T_e = 2/3 \langle u \rangle$ ($\langle u \rangle$ being the mean electron energy obtained by the actual EEDF):

$$T_g \ll T_e.$$

Usually for these plasmas $T_e \sim$ 1–2 eV, and $T_g \sim$ 1000–4000 K. In our case we have produced Argon plasma torch with $T_g$ close to the room temperature. In this way the heating effect of the plasma is eliminated so that the plasma torch can be touched (figure 3) and can be used for treatment of temperature sensitive materials and even of living tissues.

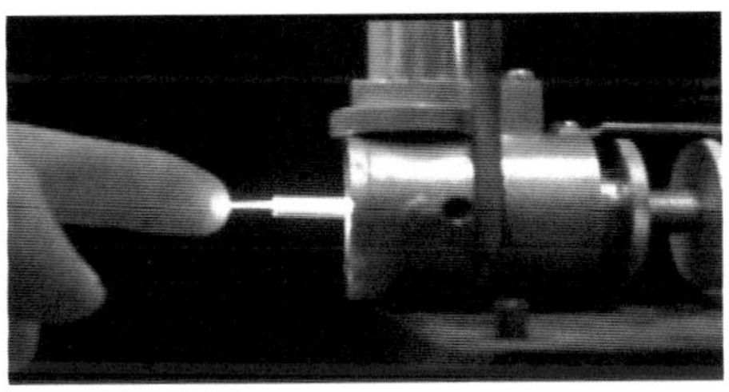

**Figure 3.** Plasma torch at atmospheric pressure at $T_g$ close to room temperature.

## 2. Experimental investigation and results

### *2.1. Physicochemical plasma characteristics of plasma torch in contact with water*

Two aspects of the plasma–liquid interaction have been investigated – on one hand the changes of the plasma characteristics and on the other the modification of the water properties. The plasma torch is considered as plasma–air configuration. Working gas is Argon at atmospheric pressure. Discharge conditions are: plasma radius 0.05 cm, 0.1 cm and 0.15 cm sustained by surface wave of frequency 2.45 GHz; gas flow rate varying from 0.1 l/min up to 1 l/min. Treatment time during all the investigations is significantly short – less than a minute and the input wave power is in the range of 12 W to 40 W. The low gas temperature allows plasma to be in close contact to the treated water.

In the presence of water placed below the discharge the geometrical dimensions of the plasma change. They have been measured and compared to those without water at the same discharge conditions (wave power, gas flow and plasma radius).

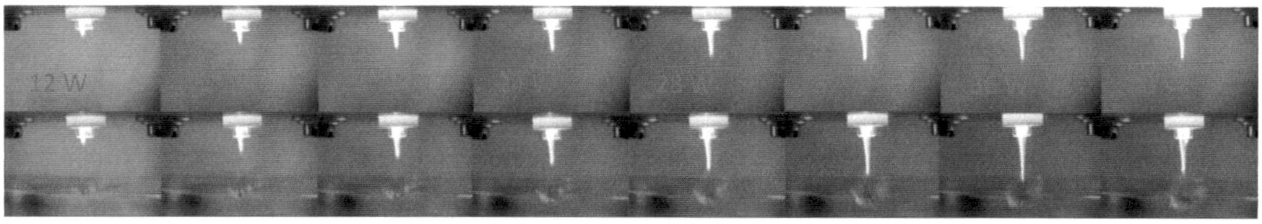

**Figure 4.** Plasma column length at internal tube diameter 3 mm and various input powers. Upper row – without water; lower row – with water below the torch.

In figure 4 the discharge in air without water at different input power (upper row) is put together with the same discharge when water is present at a fixed distance at the end of the discharge tube (lower row). In figure 5 the dependence of the plasma column length on the wave power is presented. This experiment confirms the well-known almost linear dependence of the SWD length on the wave power. New and surprising result is that when the water is present below the plasma torch its length is higher. For the tube with diameter 3 mm for all the discharge configurations the plasma torch is longer especially at higher power when water is placed bellow the plasma.

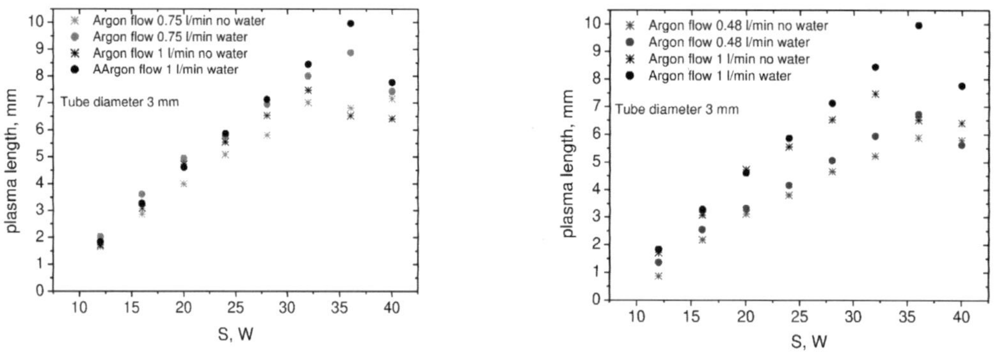


(a) (b)

**Figure 5.** Plasma column length as a function of the wave power at tube diameter 3 mm and different discharge conditions with and without water below the torch.

Estimation of the changes in the characteristic concentration of hydrogen peroxide in plasma treated liquids is done. The hydrogen peroxide concentrations have been measured in distilled water being in direct contact with the plasma. The surface-wave Argon plasma torch is produced in discharge tube with radius of 0.15 cm with gas flow 1 l/min. The characteristic concentration of hydrogen peroxide for different input power and different volume of treated liquid is measured immediately after the plasma treatment and 10 minutes after that in order to determinethe time dependence of the presence of hydrogen peroxide. The $H_2O_2$ concentration at two treatment times (30 s and 60 s) and different wave power is presented in figure 6. As one can expect at higher wave power the $H_2O_2$ production is much higher even at shorter time (30 s) than at smaller wave power (figure 6, left). One can see that at smaller treated water volume the $H_2O_2$ concentration is higher (figure 6, right). This can be due to the fact that the water is not mixed during the treatment process but only the water surface is in contact with the plasma.

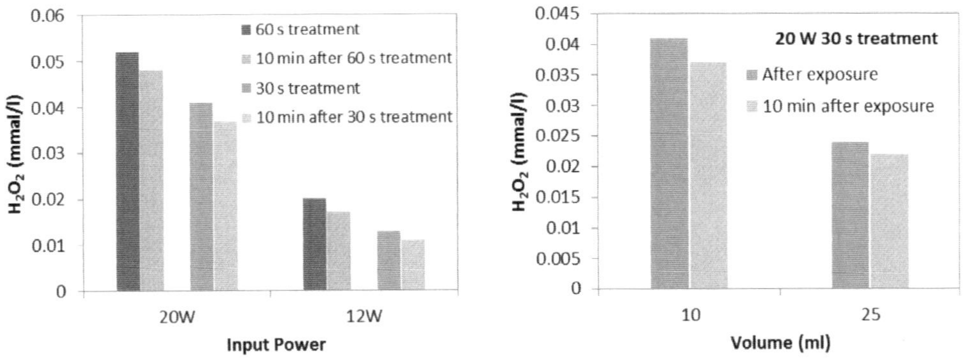


**Figure 6.** Concentration of $H_2O_2$ at different wave power, treatment time and water volume.

The hydrogen peroxide plays an important role in various biological, biomedical, environmental and wastewater treatments. It is one of the most powerful oxidizers, able to destroy organic contaminates which are ordinarily difficult to destroy, as well as inactivating cells of living organisms. This is the reason why itscreation and concentration determination is so important.

### *2.2. SWD for bio-medical applications*

We have experimentally tested the efficiency of nonthermal atmospheric pressure plasma torch in treatment of bacterial suspensions.The experimental set-up is shown in figure 7.

**Figure 7.** SWD treatment of bacterial suspension.

Only the surface of the suspension is treated by the plasma since at the low power used the plasma cannot penetrate inside the liquid and the liquid is not mixed during the treatment. The discharge conditions and the types of bacteria in the suspension are shown in the scheme in figure 8. The wave power is 20 W and the treatment time does not exceed 1 min.

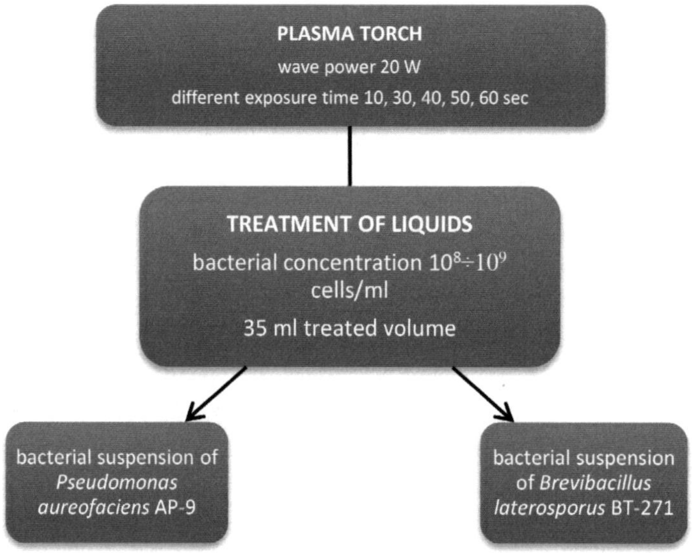


**Figure 8.** Design of bio experiments.

Two model bacterial strains *Pseudomonas aureofaciens* AP-9 and *Brevibacillus laterosporus* BT-271 were used in experiments. *P. aureofaciens* AP-9 (current classification and phylogenetic affiliation – *P. chlororaphis*) is aerobic, Gram-negative, motile, polar-flagellated, rod-shaped bacterium. *Brevibacillus laterosporus* BT-271 is a rod-shaped, Gram-positive, endospore-forming bacterium. Gram-negative bacteria are considered to be more susceptible to plasma treatment than Gram-positive bacteria because of specific cell structure and composition [3–5]. Formation of endospores as extreme survival strategy of bacteria is another factor that can affect the end results at plasma treatment. The both strains were isolated from contaminated soils and had high proven potential for biodegradation of aromatic compounds [6]. This implies their possible application as special preparations for improvement of wastewater treatment processes.

Freeze-dried bacterial cultures were rehydrated and enriched in Nutrient Broth for achievement of average initial cell concentration of $10^8$–$10^9$ cells/ml. The cell concentration was estimated by direct microscopic counting of bacterial cells in Bürker chamber (simple manual method for enumeration of average cell number by counting of cells in 80 squares – each with area 1/400 $mm^2$ and volume of sample 1/4 000 000 $cm^3$).

Plasma treatments are carried out in 35 ml volume of bacterial suspensions with different exposure time. Then bacterial suspensions were serially diluted and dispersed into the Nutrient agar in Petri dishes. The percentage of kill and log reductions were quantified by count of surviving colony forming units (CFU) after incubation at 37°C for 24 h. Each treatment and microbial analysis was repeated two or three times for reproducibility.

The results show a decreasing number of surviving cells with increase of treatment duration (figure 9). Figure 10 presents the logarithmic cell number reduction of the two bacterial strains as a function of plasma treatment time. A clear bactericidal effect of plasma treatment is observed. This effect depends on the treatment time but the type of bacteria also plays a specific role. For *Pseudomonas aureofaciens* AP-9 the inactivation effect is significant for short treatment time but the increasing of treatment time does not change it (the red columns in figure 10). The inactivation of the *Brevibacillus laterosporus* BT-271 depends stronger on the treatment time (the blue columns in figure 10). At this large treated volume (35 ml), short treatment times, and bacterial suspension with high cell density of $10^8$–$10^9$ cells/ml only partial disinfection was registered, equal to 92-99 % inactivation of bacteria.

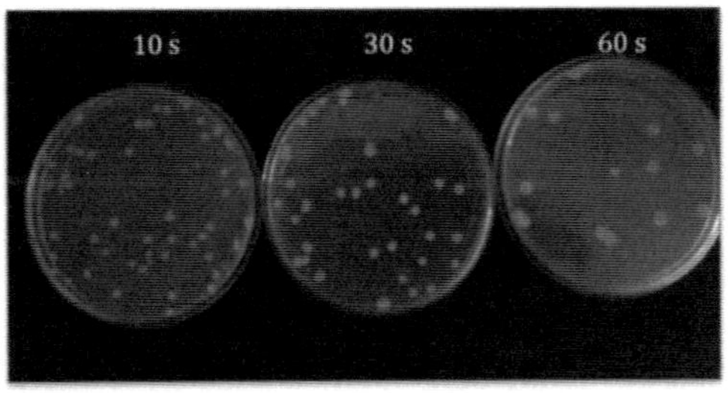


**Figure 9.** Survived colonies *Brevibacillus laterosporus BT-271*at 10, 30 and 60 sec treatment of bacterial suspension with average initial number of $7.6\times10^8$ cells/ml.

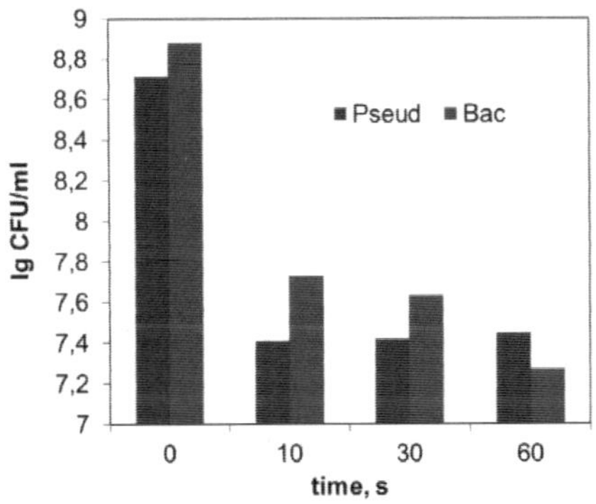


**Figure 10.** CFU reduction curves for *Pseudomonas aureofaciens* AP-9 and *Brevibacillus laterosporus* BT 271 at 10, 30 and 60 sec treatment.

Results from SEM micrographs provide a proof for plasma treatment effect on the cell level (figure 11). The morphology of pseudomonad cells before and after plasma treatment is shown in the SEM micrographs, 15 000 magnification. The untreated cells of *Pseudomonas aureofaciens* AP-9 are ellipsoidal with average dimensions of 2 µm x 0.4 µm. At the 40 sec treatment, some cells are deformed; the cell surfaces are damaged and ruptured. With increase of plasma treatment duration, the deformations and damages of cells became more serious, a mass of debris and leakage is observed.

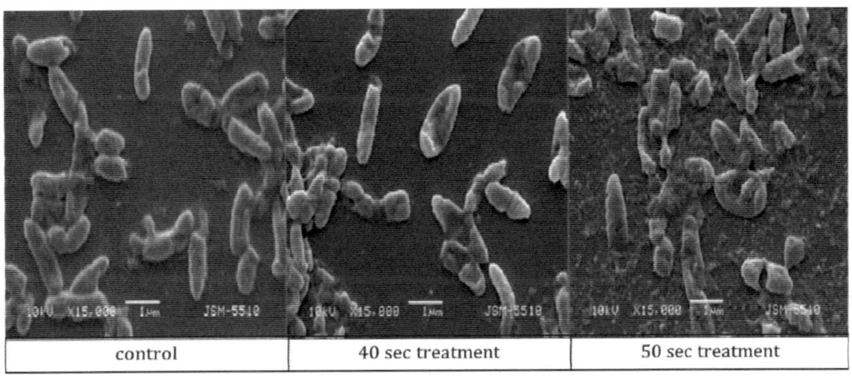


**Figure 11.** SEM pictures (scanning electron microscope JEOL JSM 5510) of: control of *Pseudomonas aureofaciens* AP-9 without plasma treatment (left), with 40 s treatment (middle) and with 50 s treatment (right).

## 3. Conclusion

In this study it has been observed that SWD plasma is able to efficiently inactivate bacteria at short exposure time less than 60 sec. The effect depends on the treatment time, volume of treated suspension and cell concentration. Although the suspension was not mixed during the treatment, the effect is noticeable. For *Pseudomonas aureofaciens* AP-9 the inactivation effect is significant for short treatment time but the increasing of treatment time does not change it. The inactivation of the *Brevibacillus laterosporus* BT-271 depends stronger on the treatment time.

After further investigations the effects of plasma on liquids can be applied for the water treatment in different facilities as a sufficient method for bacteria inactivation. The wide use of different microbial inoculants with expressed special characteristics (for example high biodegradation potential) as amendments in treatment of waters is a factor for increase of risk level from their survival in water. The plasma presents new opportunities for the better control of this unacceptable bacterial survival. The innovation also will be especially useful for the treatment of waters and materials for medical application. SWD plasma has the potential to be a cost effective, convenient and efficient method of sterilization.

**Acknowledgments**

This work was supported by Bulgarian National Science Fund under Grant No DCOST 01/03/2016 and partially by the Fund for Scientific Research of Sofia University under Grant No 80-10-204/2017.

**References**

[1] Yu H, Perni S, Shi JJ, Wang DZ, Kong MG and Shama G 2006 *J. Appl. Microbiol.* **101** 1323–30

[2] Ehlbeck J, Schnabel U, Polak M, Winter J, von Woedtke Th, Brandenburg R, von dem Hagen T and Weltmann K-D 2011 *J. Phys. D: Appl. Phys.* **44**(1) 13002

[3] Ziuzina D, Boehm D, Patil S, Cullen PJ, Bourke P 2015 *PLoS ONE* **10**(9) e0138209. doi:10.1371/journal.pone.0138209

[4] Kong M G, Kroesen G, Morfill G, Nosenko T, Shimizu T, van Dijk J and Zimmermann J L 2009 *New J. Phys.* **11** 115012

[5] Laroussi M, Mendis DA and Rosenberg M 2003 *New J. Phys.* **5** doi: 10.1088/1367-2630/5/1/341

[6] Topalova Y 2009 *Biological control and management of water treatment* (Sofia - Publish Sci-Set Eco)

[7] E. Benova, M. Atanasova, T. Bogdanov, P. Marinova, F. Krcma, V. Mazankova Microwave plasma torch at water surface 2016 *Plasma Medicine* **6**(1) 59–65

# Composite membranes with the hydrophobic and hydrophilic layers

**L I Kravets[1], A B Gilman[2], M Yu Yablokov[2], V A Altynov[1], V F Zagonenko[1]**

[1] Joint Institute for Nuclear Research, Flerov Laboratory of Nuclear Reactions, Joliot-Curie Str. 6, 141980 Dubna, Russia
[2] Enikolopov Institute of Synthetic Polymer Materials, Russian Academy of Sciences, Profsoyuznaya Str. 70, 117393 Moscow, Russia

E-mail: kravets@jinr.ru

**Abstract**. The synthesis and characterization of polymer composite membranes with hydrophilic porous substrate and hydrophobic top layer were studied. To prepare the composite membranes a polytetrafluoroethylene thin film has been applied by the technique of electron-beam sputtering deposition in vacuum on one side of a poly(ethylene terephthalate) track-etched membrane used as a porous substrate. It was found that polymer deposition on the membrane surface results in the creation of composite membranes that possess asymmetric conductivity in electrolyte solutions – a rectification effect similar to that of a *p-n* junction in semiconductors. This effect is caused by decreasing the pore diameter in the deposited polymer layer that results in change of the pore geometry, as well as existence in the pores of an interface between two layers with different nature.

## 1. Introduction

The research on the properties of composite membranes, in particular two-layered ones, which are characterized by the presence of two layers with different in both chemical structure and properties, has become a subject of essential effort. It is caused by a wide expansion of such systems. So, to two-layered membranes concern the biological membranes, membranes with the modified surface, ion-exchange membranes consisting of two cationite or two anionite layers of different nature as well as bipolar membranes consisting of two layers with antipolar conductivity. The main distinctive feature of two-layered membranes is the anisotropy of their structure and physicochemical properties. The diffusion flow of the salt through them and the forming diffusion potential differ from one another depending on the orientation of layers relative to the diffusion flow direction. When the membrane simultaneously is under the effect of both gradients of concentration and electric fields, the value of electric current passing through a membrane also depends on a disposition of the layers. These phenomena, which are typical for two-layered membranes, make transport properties asymmetric.

For example, studying the electrotransport properties of track-etched membrane (TM) from poly(ethylene terephthalate) (PET) modified by thiophene [1] and pyrrole [2] plasmas has been shown that such treatment leads to formation of composite membranes that possess asymmetry of conductivity in electrolyte solution – a rectification effect similar to that of a *p-n* junction in semiconductors. It is explained by a specific structure due to the presence of two layers with functional groups differing in chemical structure and properties. Indeed, the original membrane surface bears cation-exchange carboxyl groups [3], whereas the layer synthesized by plasma polymerization of pyrrole or thiophene contains anion-exchange nitrogen- or sulfur-containing groups, respectively. The contact of such layers with the support membranes gives rise to the unique property of conduction asymmetry. It is due to the presence of functional groups on the surfaces of both the base membrane and the plasma-deposited polymer layers, which determine the appearance of electric charge on them in electrolyte solutions, which leads to the asymmetry of conductivity.

At the same time, it is of great interest to develop methods for producing two-layered composite membranes in which one layer is hydrophobic and investigate the electrical transport properties of such systems. In this respect in some papers were investigated the properties of PET TM modified by plasmas of acetylene [2] and 1,1,1,2-tetrafluoroethane [4], which do not contain functional groups. However, it was found that the treatment by plasma of these compounds leads to the formation of composite membranes whose plasma-deposited layers bear carboxyl groups on the surface. The formation of oxygen-containing functional groups, such as COOH, in the polymers obtained by plasma is associated with both the presence of residual oxygen in the vacuum reaction chamber and oxidation of the samples during their exposure to air. The resulting effect of asymmetry in two-layered composite membranes of this type is primarily due to the contact of two layers having different concentrations of carboxyl groups in the surface layer.

In the present paper, we describe the synthesis and characterization of polymer composite membranes with hydrophilic porous substrate and hydrophobic top layer. To prepare the composite membranes a polytetrafluoroethylene (PTFE) thin film has been applied by the technique of electron-beam sputtering deposition in vacuum on one side of a poly(ethylene terephthalate) track-etched membrane used as a porous substrate.

## 2. Experimental details

In the present experiments, PET TM samples with a thickness of 9.5 μm and an effective pore diameter of 250 nm (pore density $2\times10^8$ $cm^{-2}$) were used. To produce the membrane, poly(ethylene terephthalate) film with a thickness of 10.0 μm (Lavsan, Russia) was irradiated by krypton positive ions, accelerated to ~3 MeV/nucleon at the cyclotron. Then the ion-irradiated film was additionally sensitized with ultraviolet irradiation with a maximum emission intensity at 310 nm. The chemical etching process was performed in an alkaline (NaOH) aqueous solution (3.0 mol/L) at 75°C for up to 7 min.

PTFE layers were applied onto membrane surface by physical vapor deposition using electron-beam sputtering of the polymer in vacuum. The electron source was a filament electron gun, which could produce beams with a current density of 5-100 A/$m^2$, an electron energy of 0.5-2 keV, and a cross-sectional area of (5-10)$\times10^{-4}$ $m^2$. The initial pressure of residual gases in the vacuum chamber was $\sim10^{-3}$ Pa; the substrate surface temperature, ~300 K; the emission current, 60 mA; and the electron energy, 1.2 keV. The target was a PTFE plate of 10 mm in thickness and 35 mm in diameter (Plastopolymer, St. Petersburg). The PTFE deposition rate was about 35 nm/min.

The characteristics of the original membrane and of membranes with the deposited polymer layer were determined by different procedures. The amount of the deposited polymer on the membrane surface was defined by the gravimetric method. The gas flow rate through the membranes was determined at an adjusted pressure drop. Gas consumption was measured by a float-type flow meter. From the obtained data, the effective pore diameter was calculated using the Hagen-Poiseuille equation. Surface topography of the membranes was studied using a NTEGRA Spectra (NT-MDT, Russia) atomic force microscope (AFM). Measurements were made in the dynamic mode to obtain a three-dimensional image; the scan area for each sample was 5×5 μm. The main parameter to be determine a surface roughness was root-mean-square deviation of the surface profile ($R_q$), calculated over all scan points. For each sample, scans of the surface at five different points were obtained and the $R_q$ value was then averaged. The surface properties were characterized by values of the contact angle measured with an Easy Drop DSA100 instrument (KRUSS, Germany) and the software Drop Shape Analysis V.1.90.0.14 using deionized water as a test liquid (error ±1°). Measurements of the current-voltage characteristics of the membranes were carried out in direct current regime, in the voltage range of –1 to +1 V, using a PC-controlled potentiostat Elins P-8S (Russia) with a scan rate of 50 mV/s. A two-chambered cell with Ag/AgCl electrodes, containing a water solution of potassium chloride (KCl) of identical concentration on both sides of the membrane was used for this purpose. The concentration of KCl was varied in the range of $10^{-4}$ to $10^{-1}$ mol/L.

The chemical composition of the deposited PTFE layer was studied by X-ray photoelectron spectroscopy (XPS) and attenuated total reflection Fourier transform infrared (ATR-FTIR) spectroscopy. The XPS spectra were recorded using Thermo Scientific K-Alpha (USA) instrument with a hemispherical analyzer. For the excitation of photoelectrons, X-ray radiation of an aluminum anode (AlKα = 1486.6 eV) generated at a tube voltage of 12 kV and at an emission current of 3 mA was used. The peaks' positions (the binding energy values) were calibrated to the C1s standard peak (284.6 eV). The FTIR reflection spectra were measured on a Thermo Nicolet Nexus 6700 infrared spectrometer in the range of 400-2000 $cm^{-1}$ with an optical resolution of 4 $cm^{-1}$ (digital resolution of 1 $cm^{-1}$), by averaging 32 scans. The assignment of absorption bands was carried out according to published data [5]. In order to record the XPS and FTIR-spectra of the PTFE obtained by electron-beam dispersion, the polymer was deposited on a silicon plate of 10×15 mm size.

## 3. Results and discussion

The characteristics measured for the original PET TM and the membranes prepared by applying a polymer layer using electron-beam sputter deposition of PTFE are shown in the table 1, where all results are expressed as means standard deviations of values from triplicate experiments. From these data, it follows that the deposition of the PTFE layer on the PET TM surface leads to a gradual increase in mass of the sample with the increasing time of the process. The thickness of the membrane increases and the effective pore diameter decreases in this case. Thus, during its deposition for 4.5 min, the thickness of applied PTFE layer reaches 150 nm and the effective pore diameter decreases by 55 nm. In the case of membrane treatment for 7 min, the thickness of the deposited PTFE layer is 250 nm and the effective pore diameter decreases by 75 nm. A decrease in the effective pore diameter of the composite membranes means that the pore diameter in the deposited polymer layer is much smaller than that of the original PET TM.

**Table 1.** Change of the membrane characteristics by applying a PTFE layer on the surface.

| Parameters | Original membrane | Modified membrane (min) | |
|---|---|---|---|
| | | 4.5 | 7.0 |
| Relative increase in the mass (%) | – | 2.9 ± 0.1 | 4.2 ± 0.1 |
| Thickness of deposited layer (nm) | – | 150 ± 5 | 250 ± 5 |
| Air flow rate at $\Delta P=2\times10^4$ Pa (ml/min $cm^2$) | 300 ± 5 | 125 ± 3 | 85 ± 3 |
| Effective pore diameter (nm) | 250 ± 7 | 195 ± 5 | 175 ± 5 |
| Water contact angle (deg) | 65 ± 1 | 115 ± 1 | 140 ± 1 |
| Roughness, $R_q$ (nm) | 2.1 ± 0.2 | 13.1 ± 1.2 | 30.5 ± 2.3 |

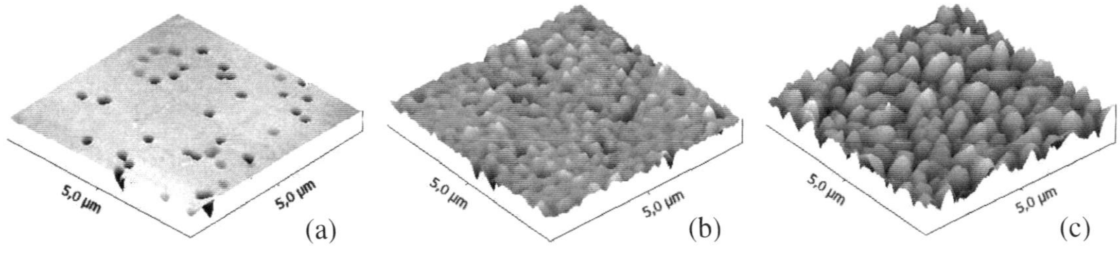


**Figure 1.** 3D AFM-images of the surface of the original PET TM (a) and the membranes after deposition the polymer layer by electron-beam sputtering of PTFE for 4.5 (b) and 7 min (c).

Examination of the surface of the modified membranes using AFM (figure 1) shows that the polymer deposition occurs mainly on the sample surface. For a membrane treated for 4.5 min, only an insignificant reduction of the pore diameter on the membrane surface is observed (figure 1b). The membrane pores are open in this case. In contrast, for a membrane treated for 7 min a PTFE layer on the treated side covers the membrane pores (figure 1c). Moreover, from the AFM data it follows that the deposited surface PTFE layer has globular character. The globules, constituting the porous system of the layer, are polymer nanoclusters with an irregular shape and a size ranging from 100 to 400 nm. The micro-relief of the membrane surface is changed. Characterization of the surface topography of the modified membranes using AFM shows a changing of the surface relief upon treatment. For the original membrane, the root-mean-square deviation of the surface profile $R_q$ is equal to 2.1 ± 0.2 nm. In order to remove the pore influence in the determination of surface roughness, smaller areas without pores, was measured in this case. The standard deviation of the surface profile $R_q$ for the modified membrane for 4.5 min is estimated at 13.1 ± 1.2 nm. A polymer layer formed on the PET TM surface during the deposition for 7 min has the $R_q$ value of 30.5 ± 2.3 nm. In addition, according to the AFM data, the pore diameter on the backside of the modified membrane remains unchanged. This means that the polymer is deposited only of a part of the pore surface.

**Table 2.** Atomic compositions of the membrane layers.

| Sample | Atomic concentration (at.%) | | | Atomic ratio | |
|---|---|---|---|---|---|
| | C | O | F | C/O | F/C |
| Original PET TM | 74.5 | 25.5 | – | 2.9 | – |
| Deposited PTFE layer | 35.5 | – | 64.5 | – | 1.8 |

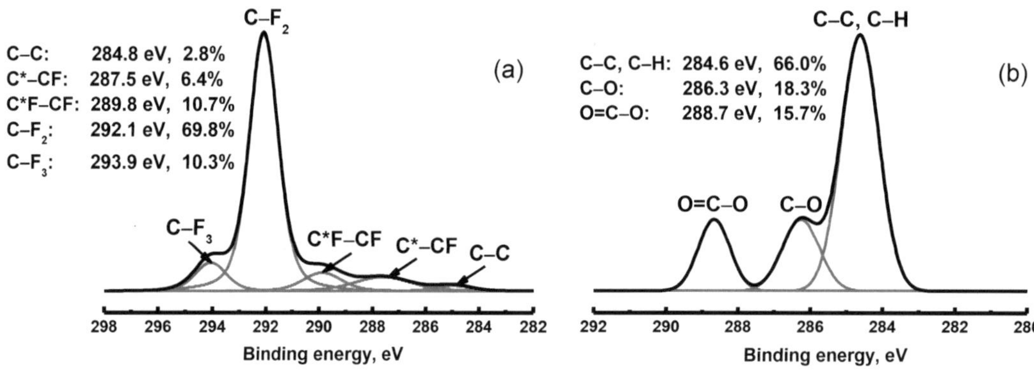


**Figure 2.** Carbon (C1s) spectra of the polymer obtained by electron-beam sputter deposition of PTFE in vacuum (a) and the original PET TM (b).

The study of the surface properties of the membranes has shown that the deposition of the PTFE film leads to substantial hydrophobization. The original membrane has the water contact angle ($\theta$) of 65°, whereas the PET TM with a PTFE layer on the surface has $\theta$ = 115° for the membrane treated for 4.5 min and 140° for the membrane treated for 7 min. An increase of contact angle can be explained by increasing the roughness in this case [6]. To chemical structure of the polymer obtained by the electron-beam sputter deposition of PTFE in vacuum was determined using XPS spectroscopic measurements. Analysis of the XPS spectrum of the PTFE layer shows the presence of the peaks due to carbon and fluorine atoms (table 2). Unlike the material obtained by plasma polymerization of 1,1,1,2-tetrafluoroethane, this polymer does not contain oxygen atoms. It is due to the formation of a certain amount of polar oxygen-containing groups during the plasma polymerization of 1,1,1,2-tetrafluoroethane that is responsible for a lower value of $\theta_{water}$ = 88° [4]. Theoretically, the fluorine to carbon atomic ratio in PTFE is equal to 2. In the polymer layer obtained by sputter deposition of PTFE, this ratio is 1.8, indicating the detachment of a certain amount of fluorine atoms during the deposition. A detailed analysis of the C1s spectrum (figure 2a) shows that it has a complex pattern due to the presence of C–C bonds (284.8 eV), $C-F_2$ (292.1 eV), C*–CF (287.5 eV), C*F–CF (289.8 eV) and $C-F_3$ (293.9 eV). Presence of the last testifies to presence in the polymer of macromolecules with short chains.

Examination by FTIR spectroscopy of the chemical structure of the polymer obtained by the electron-beam sputtering of PTFE in vacuum has shown resemblance of its IR spectrum (figure 3) to the well documented spectrum of starting PTFE [7]. The IR spectrum of PTFE displays the most intense bands at 1154 and 1210 $cm^{-1}$ due to symmetric and asymmetric stretching vibrations of $CF_2$ groups and a shoulder at 1233 $cm^{-1}$ due to C–C stretching. In the low-frequency range, there are a 634 $cm^{-1}$ band due to wagging vibrations of $CF_2$ groups and bands at 551 and 520 $cm^{-1}$ due to $CF_2$ bending and rocking vibrations, respectively. All of the these bands are observed in the IR spectrum of the polymer sample prepared by the electron-beam sputter deposition of PTFE, indicating similarity in structure of the polymers in question. It is known that the PTFE molecule is a zigzag chain made of $CF_2$ units and folded into a helix with an angle of deviation from the plane of 17°. The presence of dextro- and levorotatory helices in the polymer structure is manifested by bands at 640 and 622 $cm^{-1}$, and an intense band at 502 $cm^{-1}$ indicates a substantial degree of crystallinity of the sample. Similar absorption bands are observed in the spectrum of the polymer obtained by the electron-beam sputter deposition of PTFE, but their intensity is much lower. Thus, it may be assumed that the degree of crystallinity of the polymer structure is lower than that of the original PTFE. In addition, the most significant differences in spectra between the test polymers are the differences in intensity between the bands at 985 and 1786 $cm^{-1}$, which are due to symmetric stretching vibrations of $-CF_3$ groups (985 $cm^{-1}$) and stretching vibrations of $-CF=CF_2$ groups (1786 $cm^{-1}$). A measurable intensity of these bands in the spectrum of the deposited polymer indicates the presence of macromolecules with short chains in the sample. Thus, the PTFE layer deposited on the PET TM surface is mainly made from polymer chains composed of $-CF_2-$ units as the initial polytetrafluoroethylene. There are also macro molecules with substantially shorter chains, as evidenced by noticeable concentrations of the $-CF_3$ and $-CF=CF_2$ end groups.

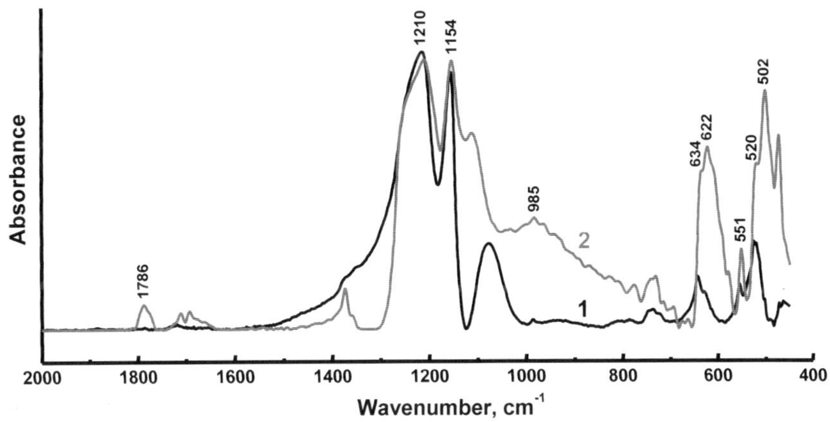


**Figure 3.** FTIR spectra of the polymer obtained by electron-beam sputter deposition of PTFE in vacuum (1) and the polymer used for sputter deposition (2).

To characterization of the chemical structure of porous substrate, the XPS spectrum was recorded also. The surface composition of the original PET TM is shown in table 2. We can see the presence of the peaks due to carbon and oxygen atoms in the XPS spectrum. Figure 2b shows the detailed structure of the carbon C1s peak for the original PET TM. The peak consists of three sub-peaks belonging to the C–C and C–H bonds (284.6 eV), to the C–O bond (286.3 eV) and to the O=C–O bond from the ester group (288.7 eV). Thus, the deposition of the PTFE layer onto the PET TM surface leads to the formation of the composite membrane consisting of two layers, one of which is the original PET matrix bearing terminal carboxyl groups on the surface and having an average level of hydrophilicity. The value of $pK_{COOH}$ for PET is 3.6-3.7; i.e., these groups undergo dissociation in a potassium chloride solution (pH 6.0). The dissociation leads to appearance of a negative charge on segments of the macromolecules, which facilitates the swelling of the membrane surface layer and the formation of a polyelectrolyte gel. The other layer deposited by electron-beam sputtering of PTFE does not contain ionizable functional groups and has hydrophobic properties.

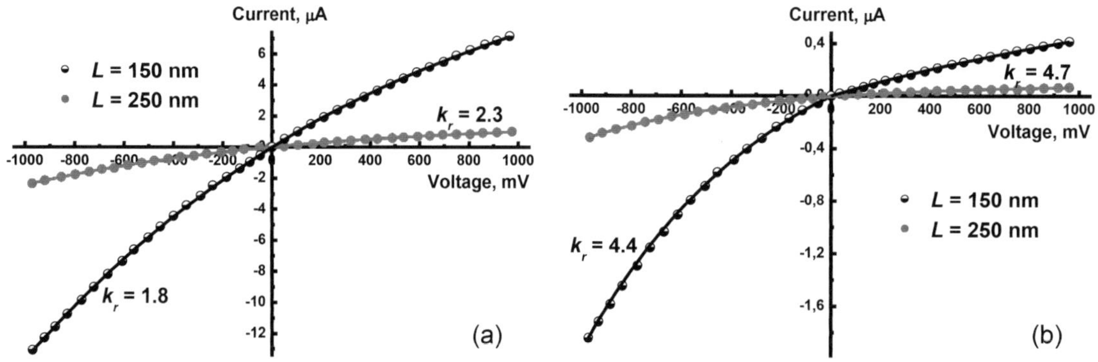


**Figure 4.** Current-voltage characteristics of the PET TM with the PTFE layers various thickness (*L*) in KCl solution with concentrations of $10^{-1}$ (a) and $10^{-2}$ mol/L (b).

The results of measurement of current-voltage characteristics for the test composite membranes show that its conductivity depends upon the current direction (figure 4). At the same time, as shows our results [1, 2] the conductivity of the original PET TM is independent of the direction of current. This means that the deposition of a hydrophobic polymer layer on the surface of the PET track-etched membrane imparts conduction asymmetry, the rectifying effect. The magnitude of this effect is characterized by the coefficient of rectification ($k_r$), which is defined as the ratio of the current at potentials of +1 and –1 V. The research shows that the rectification coefficient for the membrane with the PTFE layer thickness of 150 nm in KCl solution with a concentration of $10^{-1}$ mol/L equals 1.8, and in the solution with a concentration $10^{-2}$ mol/L it is 4.4. For the membrane with the PTFE layer thickness of 250 nm these values are 2.3 and 4.7 accordingly.

According to our earlier studies of the properties of track-etched membranes modified by pyrrole and acetylene plasmas [2], the asymmetry effect is caused by a significant reduction in the diameter of pores in the deposited polymer layer, resulting in a change of the pore geometry. The appearance of the conduction asymmetry for these membranes can also be associated with the existence of the interface between the original membrane and the deposited polymer layer. The interface is due to the presence of functional groups of different polarities in the surface layer in the case of plasma polymerization of pyrrole or different concentrations of carboxyl groups in the case of acetylene plasma polymerization. The PTFE layer in the composite membrane based on PET TM, as our results show, does not contain functional groups. Apparently, the main cause of the conduction asymmetry in this case is a decrease in pore diameter in the polymer layer deposited in the pores and resulting in a change in their geometry: the cylindrical PET TM pores acquire an asymmetrical shape during the deposition of the PTFE layer. In our opinion, for the description of the electrochemical properties of composite membranes with hydrophobic top layer should also be taken into account the phenomenon of slippage of the electrolyte solution on the hydrophobic surface [6]. Latest experiments led to the conclusion that water can slide on a hydrophobic surface [8] thereby increasing the flow velocity in hydrophobic nanochannels. The increase is because hydrophobic surfaces have a rather complex relief and can stabilize gas bubbles in cavities of the relief. Therefore, the appearance of conduction asymmetry for PET TM with a surface layer of the polymer having hydrophobic properties can be interpreted as follows. When exposing the composite membrane in a solution of electrolyte to negative potentials under the action of the direct electric current (when the membrane surface on which the PTFE layer was deposited, was turned to the anode), a significant decrease in the resistance of the system is observed. It is caused by changing the number of the ions transferred in each layer of the membrane and appearing of a concentration polarization effect. So, $K^+$ ions in this case, due to the slippage, easily pass the hydrophobic layer of PTFE. The number of the $K^+$ ions transferred in the pores of the initial membrane, apparently, is a little bit less than the number of these ions transfers in the PTFE layer, because in the case of a smooth hydrophilic surface no slippage effect takes place. This leads to accumulation of ions in the pore volume on the boundary of the PTFE layer and PET TM. $Cl^-$ ions moving in the opposite direction are also accumulated in the pore volume on the boundary of this layer and PET TM, since the number of anions transferred through the gel phase with the anion parts is lower than in a solution of electrolyte. Increase of the local KCl concentration on the boundary of the PTFE layer and the polymer matrix of PET TM promotes a current transfer.

When exposing the composite membrane in the electrolyte solution to positive potentials (the membrane surface on which PTFE was deposited is turned to the cathode in this case) under action of direct electric current, an increase of the resistance of the system is observed. At such orientation of the composite membrane, there is a fast transfer of $K^+$ ions through the PTFE layer from the pore volume into the solution. This leads to decreasing their local concentration on the boundary of this layer and PET TM. Compensation of ions from the volume of the electrolyte solution looks difficult, since the number of $K^+$ ions in the pores of the initial membrane, apparently, is less than the number of the ions transferred in the PTFE layer in view of slippage along the hydrophobic surface. The motion of $Cl^-$ ions to the anode also leads to decreasing their concentration in the membranes pore volume on the boundary of the PTFE layer and PET TM. It causes decreasing the local concentration of $Cl^-$ ions on the boundary of PTFE layer and polymer matrix PET TM thus leading to reduction of current in the system.

## 4. Conclusion

In summary, the formation of a hydrophobic polymer layer on one side of a track-etched PET membrane by electron-beam sputter deposition of PTFE results in a composite membrane exhibiting asymmetry of conduction in an electrolyte solution, the rectification effect. This is due to both a reduction in the pore diameter in the deposited polymer layer, resulting in a change in the pore geometry, and the presence of an interface in the pores between the substrate membrane and the polymer layer, which have different wettability. In description of the electrochemical properties of membranes of this kind, the phenomenon of slippage of an electrolyte solution on the hydrophobic surface should be taken into account along with the electrostatic interactions between the electrolyte ions and the charge existing on the pore walls.

## Acknowledgement

This work was supported by the grant (No 14-08-00896) from Russian Foundation for Basic Research.

## References

[1] Kravets L, Dmitriev S, Dinescu G, Satulu V, Gilman A and Yablokov M 2010 *Mater. Sci Forum*, **636-637** 812
[2] Kravets L, Dmitriev S, Satulu V, Mitu B and Dinescu G 2014 *Rom. Rep. Phys.* **66** 1165
[3] Apel P Yu and Kravets L I 1991 *High Energy Chemistry* **25** 138
[4] Kravets L I, Dmitriev S N, Satulu V, Mitu B and Dinescu G 2011 *Surf. & Coat. Techn.* **205** 455
[5] Larkin P J *Infrared and Raman Spectroscopy: Principles and Spectral Interpretation*, Waltham: Elsevier Science, 2011
[6] Vinogradova O I and Dubov A L 2012 *Mendeleev Commun.* **22** 229
[7] Ignateva L N and Buznik V M 2009 *Russian Journal of General Chemistry* **79** 677
[8] Vinogradova O I and Yakubov G E 2006 *Phys. Rev. E* **73** 045302.

# Plasma surface modification of polypropylene track-etched membrane to improve its performance properties

**L I Kravets[1], V M Elinson[2], R G Ibragimov[3], B Mitu[4], G Dinescu[4]**

[1] Joint Institute for Nuclear Research, Flerov Laboratory of Nuclear Reactions, Joliot-Curie Str. 6, 141980 Dubna, Russia
[2] Moscow Aviation Institute (National Research University), Volokolamsk Highway 4, 125993, Moscow, Russia
[3] Kazan National Research Technological University, Karl Marks Str. 68, 420015 Kazan, Russia
[4] National Institute for Laser, Plasma and Radiation Physics, Atomistilor Str. 111, 77125 Magurele, Bucharest, Romania

E-mail: kravets@jinr.ru

**Abstract**. The surface and electrochemical properties of polypropylene track-etched membrane treated by plasma of nitrogen, air and oxygen are studied. The effect of the plasma-forming gas composition on the surface morphology is considered. It has been found that the micro-relief of the membrane surface formed under the gas-discharge etching, changes. Moreover, the effect of the non-polymerizing gas plasma leads to formation of oxygen-containing functional groups, mostly carbonyl and carboxyl. It is shown that due to the formation of polar groups on the surface and its higher roughness, the wettability of the plasma-modified membranes improves. In addition, the presence of polar groups on the membrane surface layer modifies its electrochemical properties so that conductivity of plasma-treated membranes increase.

## 1. Introduction

In the majority of cases, to impart the polarity of polymer materials including membranes ones, their surface is hydrophilized. There are various methods for hydrophilization of polymer materials based on physical and chemical effects [1]. The plasma treatment is the most widely used procedure among the physical methods of hydrophilization of the membrane surface. It is advantageous because only a thin surface layer is modified due to the small penetration depth of active plasma particles into the polymer volume. The main mass of polymer remains unchanged, which is very important as regards preservation of mechanical and physicochemical properties of materials modified. In this respect, several research works published recently reported on the plasma effect on polymer membranes, including the track-etched ones, with the purpose of modifying their properties [2, 3]. It was shown that the low-temperature plasma formed by gases both non-polymerizing and polymerizing under the discharge conditions on membranes of different types induces changes in quite a number of properties: adsorption, selectivity and transport properties. Studying these processes as a function of discharge parameters and the plasma-forming gas composition with the aim of improving the membrane properties is dictated by both practical and scientific problems. In connection with this, of great interest is studying the surface hydrophilization for polypropylene track-etched membranes (PP TM) the hydrophobic properties of which prevent their use in microfiltration of aqueous solutions.

The present work focuses on the investigation of surface and electrotransport properties of PP TM modified by plasma of non-polymerizing gases in the attempt of determination of a quantitative dependence between the concentration of functional polar groups in the surface layer of membranes and their electrotransport properties.

## 2. Experimental details

In experiments, we used PP TM with an effective diameter of 250 nm that was prepared from polypropylene foil Torayfan T2372 (Toray Co., Japan) with a thickness of 10.0 μm. According to the manufacturer, this foil has biaxial orientation and consists of isotactic polymer. The foil material density was 0.92 g/cm$^3$, which corresponds to the degree of crystallinity of 80%. To prepare a membrane, the PP-foil was irradiated by xenon positive ions accelerated up to energy of ~1 MeV/nucleon at the cyclotron. The ion fluence was $10^8$ cm$^{-2}$. In order to obtain hole pores, the chemical etching of irradiated foil was carried out at a temperature of 80°C in a chromium (VI) oxide water solution with concentration 1000 g/L. The etching procedure is described in detail elsewhere [4].

The plasma treatment of membrane was carried out on a plasma-chemical facility in radio-frequency discharge (13.56 MHz) at a gas pressure of 22.5 Pa and the discharge power of 400 W in the dynamic mode. Before supplying the plasma-forming gas, the vacuum chamber was evacuated up to the residual pressure of $10^{-2}$ Pa. Membrane samples of the size of 10×10 cm$^2$ fixed on a planar holder were placed in the plasma zone of the vacuum chamber. Both sides of the membrane were plasma modified in turn. The total time of exposure to plasma was 3 min. Nitrogen, air and oxygen were used as plasma-forming gases. The scheme of plasma-chemical facility and the procedure of treatment are described in detail in [5].

The characteristics of original and plasma-modified membranes were determined by the following procedures. The variations in the membrane thickness were measured by means of a thickness meter Tesa Unit (Austria); the accuracy of measurements was ±0.1 μm. The gas permeability of membranes (for air flow through membrane) was measured at a fixed pressure difference by a float-type flowmeter. The effective pore diameter was calculated based on the acquired information by using the Hagen-Poiseuille equation. The membrane surface topography was examined by means of a Park System atomic-force microscope, XE 1000 model. To obtain three-dimensional images the measurements were carried out in non-contact mode. The scanned field was 10×10 μm$^2$ for each sample. The main parameter sought was the surface roughness (the mean-square deviation of the surface profile $R_q$ calculated for all scanned points). For each sample, the surface was scanned in five points after which the $R_q$ values were averaged.

The surface properties were characterized by values of the contact angle measured with an Easy Drop DSA100 instrument (KRUSS, Germany) and the software Drop Shape Analysis V.1.90.0.14 using deionized water as a test liquid (error ±1°). The chemical structure of the membrane surface was studied by X-ray photoelectron spectroscopy (XPS). XPS spectra were recorded by means of a Riber LAS-3000 spectrometer equipped with an OPX-150 retarding-potential hemispherical analyzer. Photoelectrons were excited with aluminum anode X-ray radiation (Al$K\alpha$ = 1486.6 eV) at a tube voltage of 12 kV and an emission current of 20 mA. Peak positions were calibrated according to the standard C1s peak (284.6 eV) [6].

The electrochemical properties of membranes were studied by the method of voltammetry. Current-voltage characteristics of the membranes were recorded at direct current in the range from –1 to +1 V by means of a Elins P-8S potentiostat (Elins, Russia) controlled by a computer. The scanning rate was 50 mV/s. A two-chambered cell with Ag/AgCl electrodes, containing a water solution of potassium chloride of identical concentration ($10^{-2}$ mol/L) on both sides of the membrane was used for this purpose. The volume of each chamber was 2.5 mL, the membrane surface was 0.5 cm$^2$ and the electrodes were immersed in electrolyte solution in each chamber at a distance of 7 mm from the membrane surface. To suppress the electrode polarization, electrodes with large surfaces were used (~3 cm$^2$); for this purpose, the silver wire 12.5 cm long and 0.7 mm in diameter was rolled into a coil. Prior to measurements, the samples were kept in electrolyte solution for 30 min.

IOP Conf. Series: Journal of Physics: Conf. Series **982** (2018) 012011 doi:10.1088/1742-6596/982/1/012011

## 3. Results and discussion

The characteristics of membranes after the treatment by plasma of the gases under study are presented in table 1 where all results are expressed as means standard deviations of values from triplicate experiments. The analysis of experimental data shows an etching of the membrane surface layer, process proved by membrane thickness reduction and increasing of the effective pore diameter. The etching effect of the membrane pores by the discharge is illustrated in figure 1 by AFM images that represent the surfaces of the original PP TM and the plasma-treated membranes (b-d). It is evident that the pore diameter on the surface of modified membranes is larger as compared with the control (original) sample. As follows from the experiments, the rate of the surface layer etching depends on the composition of the plasma-forming gas. Plasma etching in the atmosphere of air or oxygen resulted in a gain in the effective pore diameter by 13 and 42%, respectively, as compared with the nitrogen atmosphere. This effect was observed earlier for poly(ethylene terephthalate) track-etched membrane modified by plasma [7]. Furthermore, the plasma treatment affected the surface relief. Thus, for the original membrane, the mean-square deviation of the surface profile was 50.7 ± 3.3 nm. Inducing insignificant morphological changes, the nitrogen-plasma treatment lowered $R_q$ down to 42.7 ± 3.2 nm. Thus, it this case, the surface structural heterogeneities were smoothed. The treatment by air and oxygen plasma induced considerable morphological changes in the membrane surface. On the polymer surface, numerous craters appeared, i.e., oxidized-disrupt regions, the emergence of which was explained by the difference in etching rates between amorphous and crystalline polymer regions. The membrane surface became rough. The assessed mean-square deviation of the surface profile is as follows. For air-plasma modified membrane, $R_q$ = 78.6 ± 5.2 nm; for oxygen-plasma treatment, $R_q$ increased to 196 ± 12 nm.

**Table 1.** Change of the membrane characteristics after plasma treatment.

| Parameters | Original membrane | Plasma-forming gas | | |
|---|---|---|---|---|
| | | Nitrogen | Air | Oxygen |
| Thickness (μm) | 10.0 ± 0.1 | 9.9 ± 0.1 | 9.8 ± 0.1 | 9.7 ± 0.1 |
| Air flow rate at ΔP = $10^4$ Pa (ml/min cm$^2$) | 130 ± 3 | 185 ± 3 | 270 ± 4 | 600 ± 10 |
| Effective pore diameter (nm) | 250 ± 5 | 265 ± 5 | 300 ± 6 | 375 ± 7 |
| Porosity (%) | 4.9 ± 0.2 | 5.5 ± 0.3 | 7.1 ± 0.4 | 11.0 ± 0.5 |
| Water contact angle (deg) | 120 ± 1 | 75 ± 1 | 65 ± 1 | 60 ± 1 |
| Current ($J_t$) at the potential +1 V (μA)[a] | 40 ± 2 | 80 ± 3 | 300 ± 5 | 355 ± 5 |

[a] Was measured in water solution of KCl with concentration of $10^{-2}$ mol/L.

Studying surface properties of modified membranes has shown that plasma treatment improves the surface wettability. Thus, the original membrane is characterized by the water contact angle (θ) equal to 120°; for plasma-modified membranes θ value is equal to 75° (nitrogen), 65° (air) and 60° (oxygen). Insofar as according to AFM data, the plasma treatment increases surface roughness, the improved wettability was apparently due to formation of polar functional groups on the surface. Otherwise, the wettability of modified membranes would deteriorate because for hydrophobic polymers the development of roughness favors surface hydrophobization [8]. To reveal plasma-induced changes in the chemical structure of the membrane surface layer, XPS spectra were measured. The analysis of surface layer spectra for original PP TM and plasma-modified membranes revealed the presence of peaks associated with carbon and oxygen atoms and an insignificant amount of nitrogen (table 2). On the unmodified membrane surface, chromium atoms were also found. The presence of oxygen atoms in the original membrane composition is apparently due to the chemical oxidation of the polymer surface layer because strong oxidants, such as solutions containing hexavalent chromium compounds, were used as the etching agents during manufacturing of membranes [4]. The presence of chromium atoms is also associated with chemical etching, which induces the chemisorption of chromium compounds on the membrane surface. The presence of nitrogen atoms on the surface of the original membrane can apparently be attributed to the fillers involved in the polymer foil composition. An insignificant rise in the relative content of nitrogen in membranes modified by nitrogen and air plasmas suggests that nitrogen was bound by surface macromolecules during the plasma treatment of membranes.

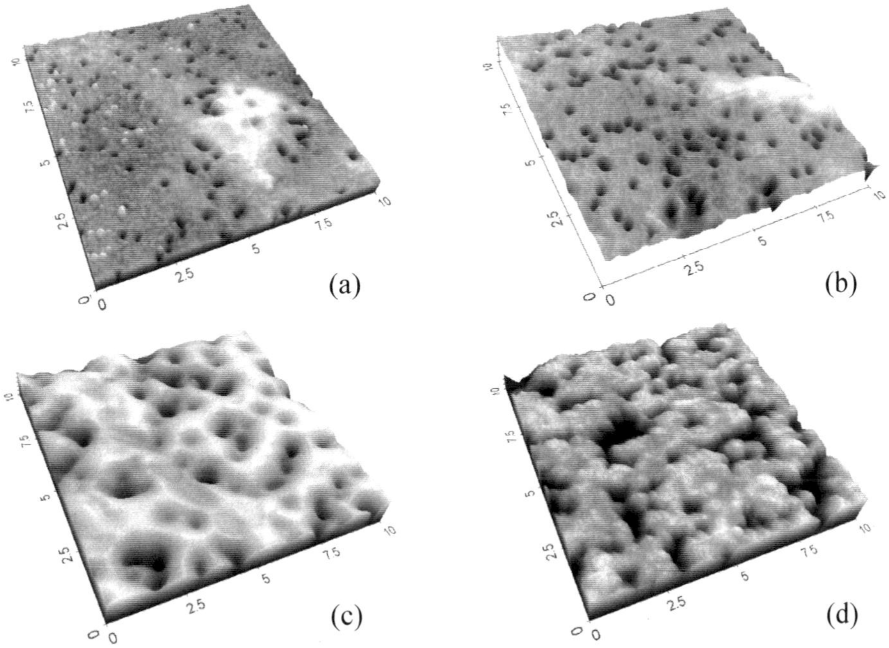


**Figure 1.** 3D AFM-images of the surface of the original PP TM (a) and plasma-treated membranes by nitrogen (b), air (c) and oxygen (d).

**Table 2.** XPS data for original PP TM and plasma-modified membranes.

| Plasma-forming gas | Atomic concentration (at.%) | | | | Atomic ratio × $10^2$ | |
|---|---|---|---|---|---|---|
| | C | O | N | Cr | O/C | N/C |
| No plasma | 93.2 ± 2.5 | 5.7 ± 0.3 | 0.80 ± 0.02 | 0.30 ± 0.01 | 6.1 | 0.85 |
| Nitrogen | 87.0 ± 2.3 | 12.1 ± 0.5 | 0.90 ± 0.02 | – | 13.9 | 1.00 |
| Air | 75.5 ± 1.9 | 23.8 ± 0.9 | 0.70 ± 0.01 | – | 31.5 | 0.90 |
| Oxygen | 71.3 ± 1.8 | 28.1 ± 1.1 | 0.60 ± 0.01 | – | 39.4 | 0.85 |

The comparative analysis of XPS data showed that modifying of PP TM by plasma formed in non-polymerizing gases increased the oxygen content in the membrane surface layer, the effect especially pronounced for air and oxygen plasmas (table 2). At the same time, the amount of nitrogen after plasma treatment depended insignificantly on the changeover of the plasma-forming gas. Hence, the decrease in the water contact angle observed upon modification by plasma formed in non-polymerizing gases is mainly due to the formation of oxygen-containing functional groups on the polymer surface. Furthermore, no chromium was found on the surface of modified membranes, which can be explained by its removal during the plasma etching of the membrane surface layer.

The chemical assignments are in accordance with Beamson and Briggs [6] and are summarized in table 3. The analysis of the original membrane XPS spectrum revealed its complex nature associated with the presence of C–C and C–H bonds in the surface layer with the concentration of 91.0% (284.6

eV) and 9.0% of oxygen-containing functional groups, among which 0.8% are the C–O–C and C–OH groups (286.5 eV), 5.9% – carbonyl groups C=O (287.7 eV) and 2.3% are carboxyl groups O–C=O (289.3 eV). The C–O–C can be assigned to an ether; C–OH can be an alcohols; C=O can be an aldehyde or a ketone; O–C=O can be attributed to an acid or ester. The nitrogen plasma treatment leads to a smaller degree of oxidation of the surface layer of membranes, concentration of carbonyl and carboxyl groups growing up to 6.7% and 3.7%, respectively, while the concentration of C–O–C and C–OH groups remains the same. The influence of air and oxygen plasma leads to additional oxidation of the surface layer of the membranes. Concentration of the oxygen-containing functional groups in the surface layer increases up to 23.1% at the air plasma treatment and up to 34.2% at the oxygen plasma, while the concentration of C–C and C–H groups decreases accordingly. The amount of carbonyl and carboxyl groups (table 3) undergoes an important enhancement in these cases. Notice, the peaks answering the binding energy of 286.5 and 287.7 eV in spectra C1s of the membrane surface can be attributed accordingly to the nitrogen-containing groups C–N and –CONH– [6]. However, the binding energy of the C–N carbon atom is equal to 285.7 eV that some smaller than the binding energy of the C–O–C and C–OH. Also, the binding energy of the amide carbon is equal to 287.9 eV that some larger than the binding energy of a carbonyl carbon. Thus, the following conclusion can be drawn based on the XPS studies of the chemical structure of the PP TM surface layer modified by plasma formed in non-polymerizing gases. The action of plasma formed by gases under study on the membrane leads to oxidation of the surface layer, which induces the formation of oxygen-containing functional groups, mainly, carbonyl and carboxyl ones. Moreover, based on the aforesaid, it can be concluded that the increase in wettability of modified membranes is due to the increase in their surface roughness and the formation of polar functional groups.

**Table 3.** Chemical bonds in surface layer of membranes.

| Plasma-forming gas | Chemical bonds in surface layer of the membranes (%) | | | |
|---|---|---|---|---|
| | C–H/C–C | C–O–C/C–OH | C=O | O–C=O |
| No plasma | 91.0 ± 2.6 | 0.80 ± 0.04 | 5.9 ± 0.3 | 2.3 ± 0.1 |
| Nitrogen | 88.8 ± 1.9 | 0.80 ± 0.05 | 6.7 ± 0.3 | 3.7 ± 0.2 |
| Air | 76.9 ± 2.2 | 2.3 ± 0.1 | 9.9 ± 0.5 | 10.9 ± 0.5 |
| Oxygen | 65.8 ± 1.8 | 7.3 ± 0.3 | 16.8 ± 0.6 | 10.1 ± 0.4 |

It is well known that the electrotransport properties of ion-exchange membranes are determined primarily by the properties of the polymer matrix – the type of functional groups, the concentration and uniformity of their distribution. Besides, the total measured electric current $I_t$ contains two components: the surface current $I_s$ (composed of counterions) flowing through the surface layer and the bulk volume current $I_b$, composed in turn from the cations and anions contributions. As $I_b$ is proportional to the volume of the channel, i.e. membrane pore diameter and $I_s$ to the channel's surface and concentration of surface functional groups. In other words, the measured current will depends on the pore diameter and chemical structure of the pore surface – type and concentration of the fixed functional groups. Measurements of current-voltage characteristics of the membranes under study show that the modified membranes possess a higher electric conductivity. So, if the value of transmembrane current $I_t$ at potential +1 V for the original PP TM is 40 μA, for the membrane modified in nitrogen plasma this current doubles. Moreover, the value of transmembrane current at potential +1 V for the membrane processed in air plasma is equal to 300 μA, while for the oxygen plasma-treated membrane it reaches 355 μA, thus exceeding 7 and respectively 9 times that through the original PP TM. Increase of the hydrophilicity and pore diameter of the membranes after plasma treatment, thus, leads to the growth of their electrical conductivity. Indeed, if the concentration of the oxygen-containing groups on the surface of modified membranes produced by plasma treatment increases as well as pore diameter, then their conductivity grows too, as evidenced in figure 2. Moreover, a good fit of the observed data if only the concentration of polar groups (a sum of carbonyl and carboxyl groups) is taken into account (figure 2b). This testifies that in the processes of mass transfer through the membranes the surface polar groups play the most significant role.

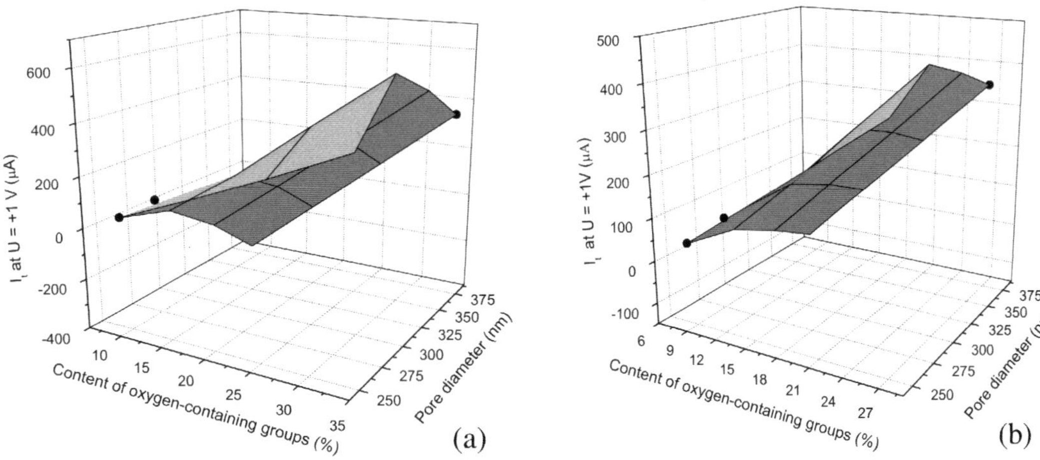


**Figure 2.** Dependence of the transmembrane current ($I_t$) at the potential $U$ = +1 V upon the pore diameter and concentration of the total content of the oxygen-containing functional groups (a) and the sum of carbonyl and carboxyl groups in the surface layer of the membranes (b).

## 4. Conclusion

Summing up the obtained results, we can make the following conclusions. When the PP TM are submitted to RF-plasma discharge in non-polymerizing gases, one can observe an etching of the membrane surface layer accompanied by reducing its thickness and increasing the effective pore diameter. Important changes in the structure and chemical composition of membrane surface layer are encountered during the gas-discharge etching. A consequence of the plasma influence is the transformation of the surface relief – the roughness of the membranes surface increases. Besides, the plasma causes oxidation of the membrane surface layer, thus leading to the formation of oxygen-containing functional groups, mainly carbonyl and carboxyl. Due to the formation of polar groups on the surface and its higher roughness, the wettability of the modified membranes improves. As a result the electrical conductivity of the treated membranes is enhanced.

## Acknowledgement

This work was performed in the frame of collaboration between the JINR, FLNR (Dubna) and NILPRP (Bucharest), Protocol no. 4429-5-15/16. We are grateful to Russian Foundation for Basic Research (grant No. 14-08-00896) for financial support.

## References

[1] Garbassi F, Morra M and Ochiello E *Polymer Surface from Physics to Technology*, New York: John Wiley & Sons, Inc., 1994
[2] Kravets L I, Dmitriev S N and Gilman A B 2009 *High Energy Chemistry* **43** 181
[3] Bryjak M, Gancarz I and Smolinska K 2010 *Adv. Colloid Interface Sci.* **161** 2
[4] Kravets L I, Dmitriev S N and Apel P Yu 1997 *Collect. Czech. Commun.* **62** 752
[5] Kravets L I, Dmitriev S N, Sleptsov V V 1998 *Nucl. Instr. and Meth. in Phys. Res. B* **142** 43
[6] Beamson G and Briggs D High Resolution XPS of Organic Polymers: The Scienta ESCA300 Database, Chichester: John Wiley & Sons, 1992
[7] Kravets L I, Dmitriev S N, Sleptsov V V and Elinson V M 2003 *Surf. Coat. Technol.* **174** 821
[8] Boinovich L B and Emelyanenko A M 2008 *Russ. Chem. Rev.* **77** 583

# Relationship of Adhesive, Contact and Electret Properties of PTFE Modified by DC Discharge

**M Yablokov, M Piskarev, A Gilman[1], A Kechek'yan and A Kuznetsov**
*Enikolopov Institute of Synthetic Polymer Materials RAS*,
ul. Profsoyuznaya 70, Moscow, 117393, Russia

E-mail: plasma@ispm.ru

**Abstract.** The relationship between the contact, adhesive and electret properties of PTFE films modified by direct current glow discharge has been studied. The film samples of 40 μm thickness were placed at the anode and cathode and treated in the air as a working gas. The contact properties of polymer surface were characterized by the values of deionized water contact angle. The peel strength was determined using T-peel test for the Scotch®810/PTFE film contact. The electret potential was measured by the compensation technique using dynamic capacitor, and from the measured potential value the effective surface charge density was calculated. It has been found that there is an undoubted correlation between the change in the value of water contact angle, the peel strength of the DC discharge-treated film, and the magnitude of the effective surface charge.

## 1. Introduction

It is well known, that the improvement of the contact and adhesive properties of polymers by low-temperature plasma treatment is usually associated with the formation of polar groups of different chemical nature, mainly, oxygen-containing, and also with structural and morphological changes due to plasma induced partial degradation of the surface layer [1–6]. However, no direct correlation between the values of the contact angle and quantitative data on the adhesion of plasma-modified polymers of different chemical nature has been revealed. Meanwhile, it has been shown that the enhancement of the contact properties of polymers may be connected with the formation of electret states in the surface layers of polymer films [7, 8]. It was found that the formation of oxygen-containing groups and the decrease in the amount of fluorine atoms on the surface of PTFE films treated by DC discharge leads to a multifold increase in the polar component of the surface energy and a marked improvement in the thermal stability of the positive charge generated by corona treatment of the plasma-processed films, probably as a result of the emergence of new deep traps on the modified surface [9]. It was established also that the stability of the positive charge significantly increases after the treatment of tetrafluoropropylene–hexafluoropropylene copolymer films by DC discharge [10]. But despite the fact that the appearance of plasma-induced electret states in polymers has been known for a long time, there is no data connecting the value of the effective surface charge to experimentally determined values of the adhesion characteristics and the contact properties of the films modified by low-temperature plasma.

For experimental measurements of the adhesion characteristics of plasma-modified films we developed a special procedure [2]. It was shown that PTFE modification by DC discharge leads to a multifold increase in the peel strength and there is an undoubted correlation between the contact properties of the polymer and the peel strength [2]. The existing electric theory of adhesion states that the adhesive properties of materials and electric charges are interrelated, and the formation of adhesive contact is represented as the contact electrification of the substrate and the adhesive [11]. The substrate and the adhesive are treated as capacitor plates, and the degradation of the adhesive contact is considered to be their separation process. However, the formation and degradation of the adhesive contact involving a charged substrate remains an open question [12].

The goal of this study is the research of the relationship of the change in the adhesion and contact properties of the surface of DC discharge-modified PTFE films with the plasma-induced formation of electret states in the polymer.

## 2. Experimental

Samples of the PTFE film of 40 μm thickness ("Plastpolymer", St. Petersburg, Russia) were used in the study. Before plasma treatment, they were conditioned by heating at 150°C in air for 30 min. The procedure for film modification by DC discharge is described in [9]. The samples were placed on the anode or cathode and treated at the air as a working plasma gas, pressure of 10–15 Pa and a discharge current of 50 mA for 60 s.

The surface properties were characterized by values of the contact angle (θ) of deionized water measured using an Easy Drop DSA100 instrument (KRUSS, Germany) with the software Drop Shape Analysis V.1.90.0.14 (error ±1°).

The electret potential was measured by the compensation technique using a dynamic capacitor [13]. The experimental setup and the procedure of measurement are detailed in [14]. From the measured value of electret potential ($U$), the effective density of the surface charge (σ) was calculated by the equation $\sigma = \varepsilon_0 \varepsilon U/L$. In this equation, $\varepsilon_0$ is the electric constant and $L$ is the thickness of the polymer film. The dielectric permittivity of the samples was taken to be $\varepsilon = 2$.

The adhesive properties of the films were investigated according to the procedure described in [2]. The peel strength ($A$) was determined using the T-peel test for the Scotch® 810/ PTFE film contact.

Values of θ, $A$, and $U$ were measured immediately after DC discharge treatment, during the storage of the films in air at room temperature and atmospheric pressure for 30 days, and after heating at 150°C for 30 min. The thermal treatment was carried out in the isothermal mode in an SNOL 58/350 oven in air at atmospheric pressure.

## 3. Results and discussion

Figure 1 (a) and (b) present data on changes in contact angle of water **θ** and peel resistance ***A***, respectively, for (1) the initial PTFE film and (2 A, C) PTFE films modified at the anode (A) and cathode (C) under the above conditions, (3 A, C) for the films after storage in air for 14 days under ambient conditions and (4 A, C) for the films after heat treatment under isothermal conditions at 150°C in air for 30 min. It can be seen that the treatment of the films at the anode is more effective than at the cathode, and the values of θ for water measured either immediately after the discharge treatment or after 14 day storage or 30 min heat treatment are lower. The films modified at the anode acquire hydrophilicity and retain this property after heat treatment or storage (θ < 60°). These films also possess better adhesion properties: the values of ***A*** measured immediately after the plasma treatment and after thermal treatment or storage were higher. These results indicate the existence of an inverse correlation between the contact angle and the peel strength: the less the value of θ, the greater the value of ***A***. This relationship is characteristic of the both types of films modified at the anode or the cathode.

Experimentally determined values of effective surface charge density (**σ**) for the PTFE films modified at the anode or cathode are shown in Fig. (c). There are data given for (1) the initial film and the films (2 A, C) immediately after plasma treatment at the anode and cathode, (3 A, C) after 14 day of storage and (4 A, C) after heat treatment. Note that the initial film had a small negative surface potential and the effective charge density of this film is σ = –15 μC/m$^2$. The presence of this potential is apparently due to the sample prehistory.

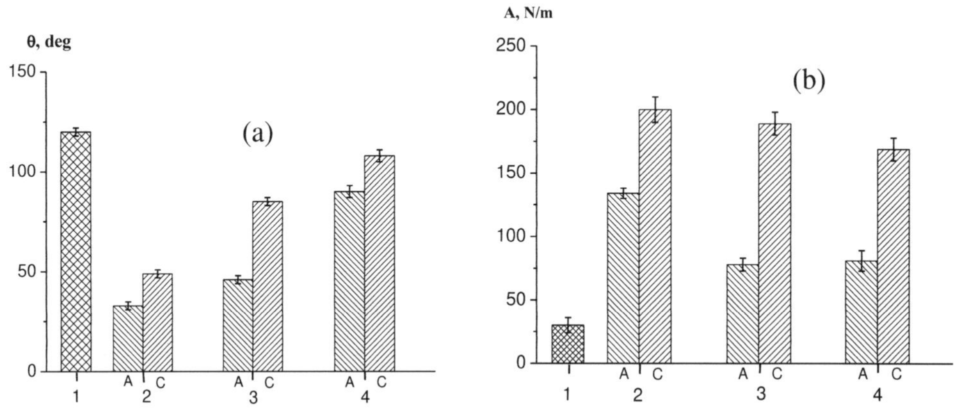


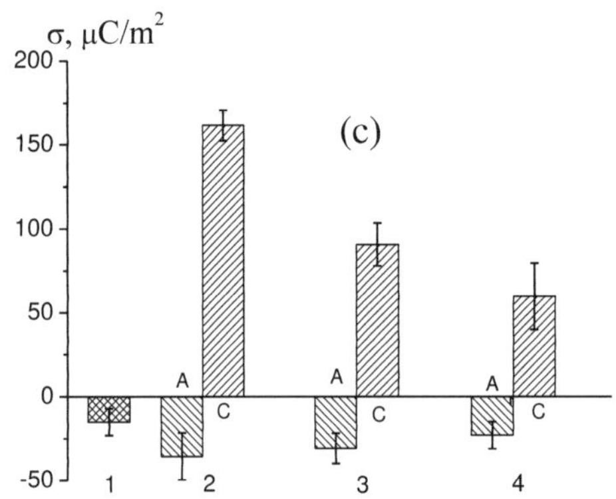


**Figure 1**. (a) values of contact angle of water **θ**, (b) of peel strength ***A*** and (c) of effective surface charge density **σ** for (*1*) the initial PTFE film, (*2*) the film modified by plasma at the anode (A) and the cathode (C), (*3*) the plasma-modified film stored for 14 days under ambient conditions, and (*4*) the plasma-modified film treated at 150°C for 30 min.

After modifying the film at the anode, the effective density of negative charge increases to –36 μC/m2, whereas the film treated at the cathode acquires a positive surface potential.

It is seen that as the storage time increases to 14 days, the contact angle increases, the peel strength decreases, and the effective surface charge density is reduced. Note that there is the undoubted correlation between the effective surface charge density and **θ**, as well as between **σ** and ***A***. These data show also that the parameters for the films treated at the anode undergo substantially smaller changes than those for the films modified at the cathode.

Thus, these data suggest that the rise of the effective density of negative surface charge in the films after their modification at the anode correlates with the increment in the peel strength, and its fall during storage of the films or after heat treatment correlates with the decrement in σ. After the treatment at the cathode, the films acquire a positive effective surface charge, and the peel strength increases as compared with the initial value, although to a lesser extent than in the case of the treatment at the anode.

It is known [13, 14] that the experimentally determined value of the effective surface charge density is an integral characteristic of the electret state and is the sum of homo- and heterocharges. During dc discharge treatment of a polymer, the homocharge is induced by injection of electrons or ions from the plasma, depending on the treatment conditions and the position of the polymer film relative to the electrodes. The heterocharge can be generated as a result of the formation of polarized states of polar groups oriented in the electric field induced in the discharge gap. These polar groups are formed in the surface layer of the polymer by chemical reactions occurring during its plasma modification. Thus, the nature of the DC discharge-induced electret state in the polymer film may be associated not only with the charges as such, but also with oriented dipoles occurring on the surface [13]. We have shown that the thickness of the PTFE layer chemically modified by plasma at the anode is 40–50 nm [15]. It is at this distance from the polymer surface that the concentration of oxygen atoms is reduced from ~10% to zero. At the same time, it is known that the homocharge is built up in a layer of ~0.1–0.01 μm in thickness [16], which corresponds to the aforementioned thickness of the modified polymer layer. The determining role in the buildup of the homocharge is played by the traps, which are associated with various defects such as terminal groups of macromolecules, long-lived free radicals, crystallite defects, the interface between the polymer crystalline and amorphous phases, and polar groups on the surface. The depth and concentration of traps in a plasma-treated polymer substantially depend on the mode of treatment [13]. Based on our experimental data, we can suggest that the treatment at the cathode gives rise to a variety of shallow traps in the surface layer of PTFE film, which are rather quickly depopulated by thermal motion. On the other hand, the PTFE film treated at the anode has a substantially larger number of polar groups on the surface [2], thereby greatly facilitating the formation of deep traps. Calculations for model systems showed [17] that polar impurities localized near structural defects can cause the formation of traps in the vicinity of the defect, themselves not being the traps. It is due to the formation of deep traps in the case of treatment at the anode that the surface potential relaxation rate for the PTFE film is lower.

In summary, the results of this study show that there is an undoubted correlation between the adhesion characteristics, the change in the contact angle, and the magnitude of effective surface charge of the PTFE film treated in DC glow discharge.

## Acknowledgments

This work was supported by the Presidential program of assistance of leading national scientific schools, grant nos. NSh-324.2012.3 and NSh-1683.2014.3.

## References

[1] *Surface Treatment of Materials for Adhesive Bonding* 2013 ed S Ebnesaijjad, C Ebnesaijjad (Norwick: William Andrew Publ)
[2] Gilman A, Piskarev M, Yablokov M, Kechek`yan A and Kuznetsov A 2014 *J. Phys.:Conf. Ser.* **516** 012012
[3] Fang Z, Hao L, Yang H, Xie X, Qui Y and Edmund K 2009 *Appl. Surf. Sci.* **255 7279**
[4] Nitschke M, Koenig U, Lappan U, Minko S, Simon F, Zschoche S and Werner C 2007. *J. Appl. Polym. Sci.* **103** 100
[5] Kusano J 2014 *J. Adhes.* **90** 755
[6] Park YW and Inagaki N 2004 *J. Appl. Polym.Sci.***93** 1012
[7] *Encyclopaedia of Low-Temperature Plasma* 2006 **11** ed. Yu Lebedev, N Plate, V Fortov (Moscow: Yanus K)
[8] Chen Q 2002 *J. Phys. D: Appl. Phys.* **35** 2939 [9] Richkov D, Yablokov M and Richkov A 2012 *Appl. Phys. A, Mater. Sci. Process.* **107** 589
[10] Richkov A, Yablokov M, Kuznetsov AE, Gilman A and Kuznetsov AA 2010 *High Energy Chem.***44** 347
[11] Deryagin BV, Krotova NA and SmilgaVP 1978 *Adhesion of Solids* (NY: Consultants Bureau)
[12] Pocius A 2002 *Adhesion and Adhesive Technology* (Munich: Carl Hanser Verlag)
[13] *Electrets* 1980 ed. GM Sessler (Berlin: Springer)
[14] Yablokov MYu, Kechek`yan AS, Gilman AB and Ozerin AN 2011 *Nanotechnika* **2(26)** 86
[15] Yablokov MYu, Sokolov IV, Malinovskaya OS, Gilman AB and Kuznetsov AA 2013 *High Ehergy Chem.***47** 32
[16] Richkov DA, Kuznetsov AE and Richkov AA 2013 *Charge Stabilization of Polymeric Electrets* (St. Petersburg: RGPU im. A.I. Gertsena)
[17] Sworakowski J 2000 *IEEE Trans. Dielectr. Electr. Insul.***7** 531

IOP Conf. Series: Journal of Physics: Conf. Series **982** (2018) 012013 doi:10.1088/1742-6596/982/1/012013

# Probe measurements of electron energy spectrum and plasma-wall interaction in Helium/air micro-plasma at atmospheric pressure

**S F Adams[1], V I Demidov[1,2], A L Hensley[3], M E Koepke[2], I P Kurlyandskaya[4], J A Miles[1] and B A Tolson[3]**
[1] Air Force Research Laboratory, WPAFB, OH 45433, USA
[2] Department of Physics, West Virginia University, Morgantown, WV 26506, USA
[3] UES, Inc., Dayton, OH 45432, USA
[4] ITMO University, Kronverkskiy pr. 49, St. Petersburg 197101, Russia

E-mail: vladimir.demidov@mail.wvu.edu

**Abstract.** It is experimentally demonstrated that a wall probe may be a useful instrument for measurement of electron energy spectrum (EES) and the plasma-boundary interaction in a micro-plasma with a nonlocal electron distribution function at atmospheric pressure. The measurements of the EES have been conducted in the plasma of several micro-hollow cathode discharges of differing sizes. The discharges with a flow of Helium gas were exposed to air. Typical results of measurements demonstrate signature of energetic electrons arising due to plasma-chemical reactions. It is experimentally shown that wall probe potential is associated with energetic electrons rather than the ambient electron kinetic energy. The devices may be applicable for developing analytical sensors for extreme environments, including high radiation and high temperatures.

## 1. Introduction

Electric probes are widely used for measurements of the electron distribution function (EDF) in a plasma [1, 2]. Cylindrical probes are normally used, where a cylindrical rod is inserted into the plasma and acts as the probe. Recently [3, 4] it was demonstrated that a wall probe might be a convenient tool for measurements of the EDF in plasmas with a nonlocal EDF [5]. The wall probe is an electrically isolated part of the volume boundary, serving to collect the current from the plasma for different probe potentials. The wall probe has benefits of greater area (higher sensitivity), absence of a probe-holder (less plasma distortion) and reduced influence of ion current (less distorted measured EDF). Due to constructive reasons, the wall probe (in contrast to a cylindrical probe) can also be more easily used in a micro-scale plasma.

For correct measurements of the EDF, the wall probe area should be much smaller than the plasma boundary area and an appropriate probe theory should be used [2–4]. In this case, the measured EDF from the wall probe would be practically equal to the EDF in the undisturbed (by the probe) plasma. In contrast, if the wall probe is large, the measured EDF (which we will refer as the electron energy spectrum (EES)) may not accurately represent the EDF in the undisturbed plasma as the plasma properties can change during the measurements. Such alternation in a plasma with nonlocal EDF can be smooth [6, 7] or sharp [8], depending on the plasma geometry. Although not ideal, a large wall probe may still be useful for applications where higher sensitivity of the probe measurements is essential, but exact knowledge of the undisturbed plasma EDF is not very important or the results of the measurements can be correctly interpreted. That is the case, for example, for the method of plasma electron spectroscopy (PLES) [2, 4], which may be useful for characterization of plasma-chemical processes in the plasma. The PLES method is based on the measurements of nonlocal energetic electrons, arising in a plasma from various volumetric plasma-chemical processes, in an afterglow or afterglow-like plasma with low-electron temperature (typically of the order of 0.1 eV) [9, 10]. Measurements of the high-energy portion of the EES (usually in the energy range between 1 and 20 eV) can provide analytical information about the gas mixture.

In this paper, we demonstrate experimentally that using the wall probe can accurately measure micro-discharge plasma EES for interpreting plasma-boundary interactions and plasma-chemical reactions in atmospheric-pressure plasma with a nonlocal energetic-electron energy spectrum. For nonlocal plasmas, the EES at the boundary is also representative of the entire plasma.

## 2. Probe theory at high-pressure plasma

A probe accurately measures the EDF of electrons within one energy relaxation length ($\lambda_\varepsilon$) from the probe. Note that for elastic collisions this relaxation length may be much greater than the mean free path ($\lambda$) since elastic collisions within the plasma will preserve electron kinetic energy. Within $\lambda_\varepsilon$, the total electron energy is not spatially dependent, and the EDF is considered nonlocal [5]. If electrons are further away from the probe (greater than $\lambda_\varepsilon$) the recorded electron total energies are now spatially dependent, and the EDF is considered local. The probe size must also be taken into consideration, as a probe larger than $\lambda_\varepsilon$ would detect different electron energies depending on the location on the probe. For a cylindrical probe the radius ($R$) determines the probe size. These parameters, $\lambda_\varepsilon$, $\lambda$, and $R$ determine how the probe measurements should be interpreted [3]. For a low pressure plasma, where $R \ll \lambda$, the Druyvesteyn formula is valid and the EDF is proportional to the second derivative of the probe current with respect to the probe potential. For a high pressure plasma, where $R \gg \lambda$ but $R \ll \lambda_\varepsilon$, the EDF is proportional to the first derivative of the probe current [1–4].

The wall probe in this work is relatively large compared to the plasma volume surface. The transport of electrons to the probe is diffusive and therefore, as mentioned above, the first derivative will be used to determine the EES. Thus, in our experiments, shown below, the measured first derivative of the probe current with respect to the applied probe potential is the experimentally recorded. The EES accurately indicates signature of plasma-chemical reactions in the plasma, but their relative ratio of secondary peaks may not correspond to the exact ratio as the undisturbed EDF in plasma. If it is necessary, for example, for determination of the undisturbed EDF, this could be corrected for by calibrating with known gas mixtures or numerical modeling so that accurate ratios can be determined. Note, that the presence of a specific target gas component can be monitored in a gas mixture from the measured EES without correction.

Thus, if $\lambda_\varepsilon$ is substantially greater than the plasma volume size, the EES can be measured by the wall probe for the entire volume. For noble gases and atmospheric pressure that corresponds to the plasma volume size of order of 100 μm. If $\lambda_\varepsilon$ is comparable or somewhat smaller than the plasma volume size, high-energy peaks at the EES still can be measured, but can be accompanied by additional background electron continuous spectrum due to partial energy relaxation of energetic electrons. Note, that adding air to Helium can reduce $\lambda_\varepsilon$ and provide the transition from nonlocal to local regime without alternation of the discharge volume or gas pressure.

## 3. Experimental devices

For the experiments, several micro-plasma short dc discharge devices were created. They were fabricated with three layers of molybdenum metal foils with thickness of 100 μm separated by two sheets of mica insulation with thickness of 110 μm. The different devices have holes with the diameter from 65 μm to 450 μm, which formed cylindrical discharge cavities that passed through the entire five layers. The holes were produced by drilling with a bit for the diameters greater than or equal to 200 μm. In the devices with the holes of less than 200 μm, the holes were produced with a laser. In all

IOP Conf. Series: Journal of Physics: Conf. Series **982** (2018) 012013 doi:10.1088/1742-6596/982/1/012013

devices, the inner molybdenum layer formed a wall probe, while the outer layers of molybdenum served as the hollow cathode and anode. The discharge was open to air with a flow of Helium gas. The pictures in figure 1 represent three different devices and plasma conditions.

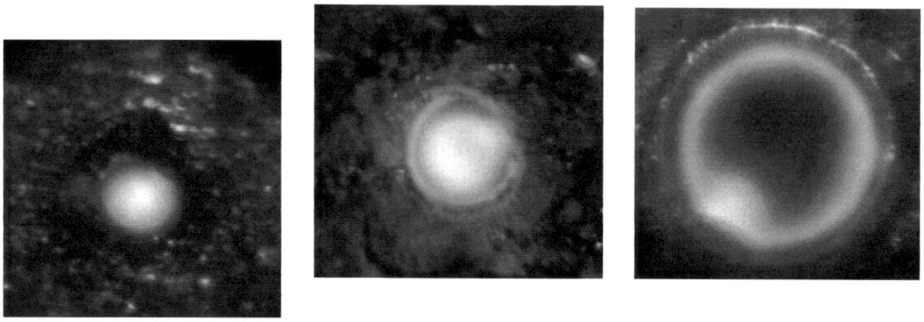

**Figure 1.** Pictures of the devices with the laser produced hole with the diameter of 100 μm (left) and the drill-bit produced hole with the diameter of 200 μm (middle) and 450 μm (right).

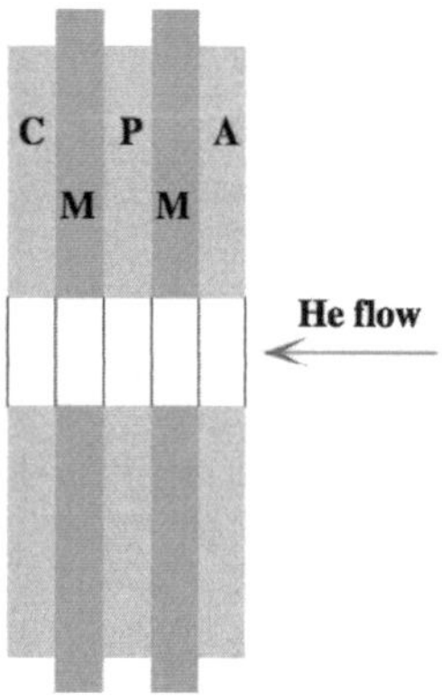


**Figure 2.** A sketch of discharge device. Hollow-cathode (C), hollow-anode (A), wall probe (P) and mica wall (M).

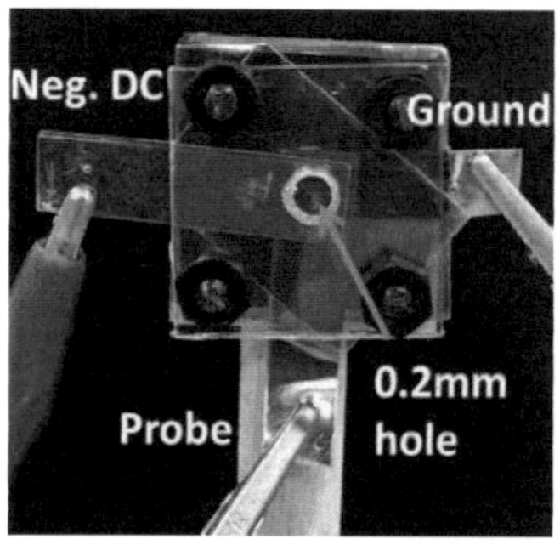


**Figure 3.** A picture of the experimental device with the hole of the diameter of 200 μm.

For the device with the 450 μm hole, the plasma is not uniform and confined to the edge of the hole. This is considered to be a local plasma. At 200 μm, the plasma is more uniform but still exhibits some non-uniformities ($\lambda_\varepsilon$ is about the discharge radius). At 100 μm and lower, the plasma is practically uniform and is assumed to be nonlocal. A sketch of the device is shown in figure 2 and a picture is shown in figure 3. EES measurements were taken with devices at 200 um and lower. It is expected that each dc device will exhibit an afterglow-like plasma, where the majority of electrons have a Maxwellian EDF with very low temperature, $T_e$, of the order of 0.1 eV. In this plasma, electrons with energies between 1 and 20 eV are created due to volumetric plasma-chemical reactions.

## 4. **Results of the experiments and discussion**

Typical results of measurements of wall probe current $j_w$ and its first derivative with respect to probe potential $dj_w/dV$ are shown in figures 4 and 5 ($I_d$ is the discharge current). It is found that the wall probe I–V trace is sensitive to the presence of energetic electrons created due to presence in the plasma of Helium metastable atoms. The first derivative of the probe current with respect to the probe potential shows peaks revealing energetic electrons at specific energies arising due to plasma chemical reactions (see figure 5). The maximums at 15 and 20 eV correspond to binary collision between He metastable atoms (Penning ionization) or de-excitation of metastable atoms by slow electrons.

Maximum about 3–5 eV corresponds to Penning ionization of Nitrogen molecules by metastable Helium atoms. The energy of created electrons depends on the condition of the Nitrogen molecule and the type of metastable atom. Groups of electrons near energies of 10 eV may correspond to Penning ionization of Oxygen by metastable Helium atoms. The energetic electron peaks at the EES for the discharge with lower diameter are more pronounce which could be connected to increasing ratio between $\lambda_\varepsilon$ and the discharge diameter. That corresponds to transition from regimes with nonlocal EES to regimes with local EES are visible in the discharges, shown in figure 1. It is seen that due to energetic electrons the wall potential can be much higher than that, associated with the ambient electron kinetic energy [11]. The device with smaller plasma diameter has lower wall potential, which is about –10 V in contrast to –29 V, as shown in figure 4.

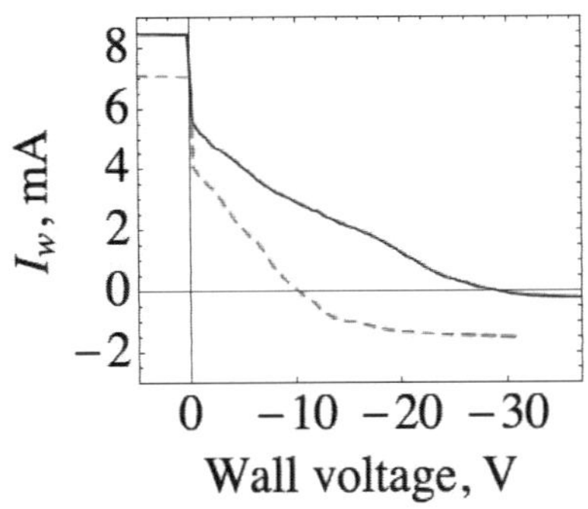


**Figure 4.** Wall current $I_w$ with respect to wall potential in a Helium dc discharge surrounded by air. Dashed curve (65-μm device, $I_d$ = 6.5 mA) and solid curve (200-μm device, $I_d$ = 8.2 mA).

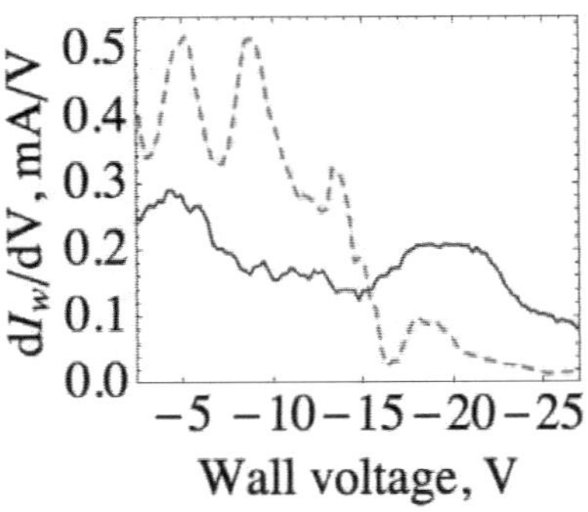


**Figure 5.** First derivative, $dI_w/dV$, with respect to wall potential for figure 5. Dashed curve (65-μm device, $I_d$ = 6.5 mA) and solid curve (200-μm device, $I_d$= 8.2 mA).

**Acknowledgments**
This research was performed, while VID held a National Research Council Research Associateship Award at AFRL. The work of IPK was supported by ITMO University (Grant No. 713577).

**References**

[1] Swift J D and Schwar M J R 1970 *Electrical Probes for Plasma Diagnostics* (London: Iliffe Books)
[2] Demidov V I, Ratynskaia S V and Rypdal K 2002 *Rev. Sci. Instrum.* **73** 3409
[3] Demidov V I and Kudryavtsev A A 2014 *Phys. Plasmas* **21** 093506
[4] Godyak V A and Demidov V I 2011 *J. Phys. D: Appl. Phys.* **44** 233001
[5] Tsendin L D 1995 *Plasma Sources Sci. Technol.* **4** 200
[6] Demidov V I, Kudryavtsev A A, Kurlyandskaya I P and Stepanova O M 2014 *Phys. Plasmas* **21** 094501
[7] Adams S F, Demidov V I, Bogdanov E A, Koepke M E, Kudryavtsev A A and Kurlyandskaya I P 2015 *Phys. Plasmas* **22** 024501
[8] Mustafaev A S, Demidov V I, Kaganovich I D, Koepke M E and Grabovsky A 2014 *Phys. Plasmas* **21** 053508
[9] Blagoev A B, Kagan Y M, Kolokolov N B and Lyagushenko R I 1974 *Sov. Phys. Tech. Phys.* **19** 215
[10] Demidov V I, Adams S F, Blessington, Koepke M E and Williamson J M 2010 *Contrib. Plasma Physics* **50** 808
[11] DeJoseph, Jr C A, Demidov V I and Kudryavtsev A A 2007 *Phys. Plasmas* **14** 057

# Time-Space Radial Development of Nanosecond Dielectric Barrier Discharge in Flat Air Gaps under Atmospheric Pressure

**E A Shershunova, V Yu Khomich, M V Malashin and S I Moshkunov**
IEE RAS, 191186, 18, Dvortsovaya nab., St.-Petersburg, Russia

E-mail: eshershunova@gmail.com

**Abstract.** The study of the time-spatial radial development of the nanosecond dielectric barrier discharge in open air was carried out by means of segmented electrode technique. It was found experimentally that the discharge development depends on the air gap height. Discharge currents through the segments in 1-mm and 2-mm air gaps started almost simultaneously with a time delay relative to each other no more than 1 ns. An increase of the delay between discharge currents appearance and a certain order of their appearance was observed with increasing the height of the gap. Different development of the discharge correlated with a different growth rate of electric field strength in the gap and the ratio *E/p*.

## 1. Introduction

Nanosecond dielectric barrier discharge (NDBD) is a discharge, characterized by a current pulse with a current density of several amperes per $cm^2$, which takes place between electrodes, if one of them is isolated from the gap by a dielectric. The NDBD can be realized by providing high growth rate of voltage at the electrodes of the discharge gap [1–4].

Obtaining the NDBD in atmospheric air at natural humidity has been a breakthrough for applied science as it allows realizing a volume low-temperature plasma without additional preionization sources in open air. Extensive research of the NDBD depending on parameters of the voltage applied to the electrodes such as amplitude, repetition rate, growth rate, has been carried out for the last ten years.

However the task of the NDBD development in millimeter air gaps is not fully understood. Evolution of the NDBD in time and space is analyzed mainly through high-speed photography with a time resolution of a few nanoseconds [5].

Nevertheless, even such method is not always possible to separate the temporal events taking place in the air gap, so another approach to the investigation of the NDBD has been used in the study. The study of the time-spatial radial dynamics of the NDBD in millimeter air gaps was conducted with the use of a ring segmented electrode [6, 7], together with the use of high-precision diagnostic devices enabled to achieve resolution accuracy of 200 ps.

## 2. Experimental setup

The experimental setup is presented in figure 1. Unipolar negative high-voltage square pulses with the pulse width of 600 ns, amplitude of 20 kV with rise time of the voltage pulse (10–90 %) ≈ 40 ns are formed at the electrodes of the discharge gap by a specially developed semiconductor high voltage generator (HVPG) [8]. Pulse repetition rate $f$ was set to 30 Hz during the experiment.

In figure 1 the discharge gap is shown as a set of capacitors C1, C2, C3 and C4, corresponding to the equivalent capacitances of the gaps, formed by the ring segmented electrode and the high voltage electrode, covered by the alumina ceramics plates with $\varepsilon = 9$ and dimensions 50×50×2 mm. Due to the same area of all the segments these capacitances were quite the same either, excepting the edge segment, as its capacity was bigger than the other ones due to the fringing field effect.

The NDBD was ignited in atmospheric air at natural humidity of 40–60 %. The height of the discharge gap ranged from 1 to 3 mm during the experiment.

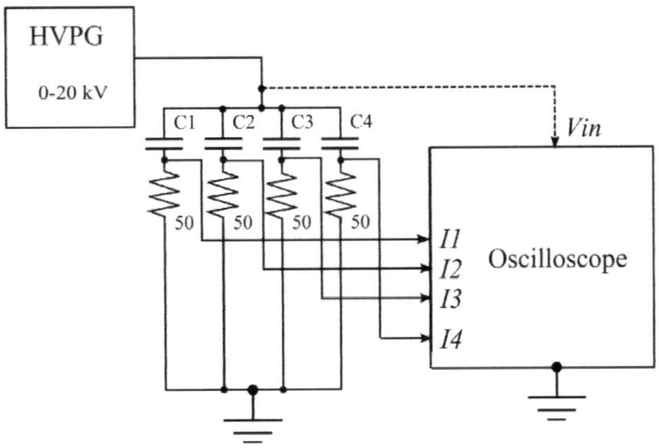


**Figure 1.** Experimental setup.

The voltage at the electrodes was measured by Tektronics P6015A high-voltage probe. The time-space development of the NDBD was estimated through the comparison of the current waveforms in different volumes of the gap I1, I2, I3 and I4, obtained using synchronized low-inductance resistive current sensors (50 Ohm). The voltage and currents signals were observed by LeCroy WaveRunner 104Xi-A oscilloscope (bandwidth of 1 GHz, sampling rate of 10 GS/s). The degree of NDBD homogeneity was analyzed through the method of optical diagnostics, which was based on comparison of the light emission brightness distribution in the gap with the brightness curve of the ideal isotropic glow source discussed in detail in [9].

## 3. Experimental results and discussion

Due to the obtained experimental results, the different delays between discharge ignitions in different segments were observed in dependence on the height of the discharge gap $d$. Currents in the segments (corresponding in figure 1: C1 – central segment, C2 – next to the central segment, C3 – next to C3, C4 – the edge segment) developed nearly at the same time at $d$ = 1–2 mm: the delay time between them was no more than 1 ns. Moreover, they appeared in a random order. The typical current waveforms in the segments at $d$ = 2 mm are shown in figure 2a.

As for $d$ = 3 mm the delay between discharges increased to 3-5 ns, and the discharges ignited in a definite way. The first discharge started in the central segment, the second – in the second segment and so on, spreading radially, as it was shown in figure 2b.

Figure 3 shows brightness distribution of the glow in the central layer of the gap cross section (layer thickness = 10 μm), obtained by the use of digital photo processing for $d$ = 1mm (figure 3a), $d$ = 2 mm (figure 3b) and $d$ = 3 mm (figure 3c).

In figure 3 value $\sigma$ is a standard deviation and it is obtained by the formula:

$$\sigma = \sqrt{\frac{1}{n}\sum_{i=0}^{n}(D_i - L(\alpha, i))^2}, \quad (1)$$

where $D_i$ – experimental brightness values, $L(\alpha,i)$ – the ideal brightness curve, obtained according to the electrode shape, $i$ – a pixel number of the photo image, corresponding to the horizontal coordinate of the gap from one edge to another, $n$ – a total number of pixels. $\xi$ is an average experimental brightness value. As one can see from figure 3 the ratio $\sigma/\xi$ is different for different heights of the gap.

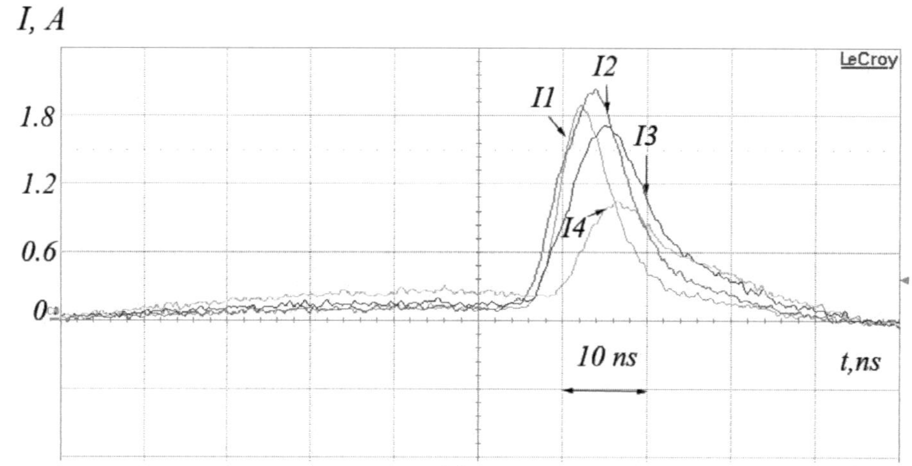


(a)

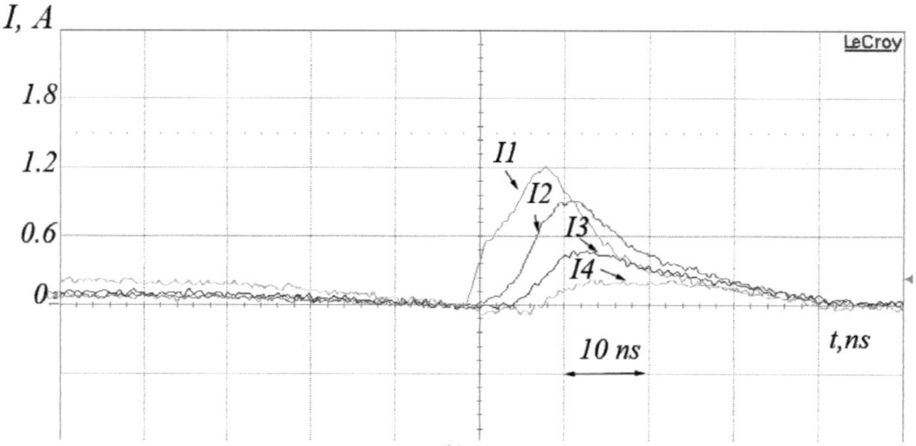


(b)

**Figure 2.** Typical current waveforms at voltage rise in different segments at various height of the discharge gap $d$: a) $d = 2$ mm, b) $d = 3$ mm

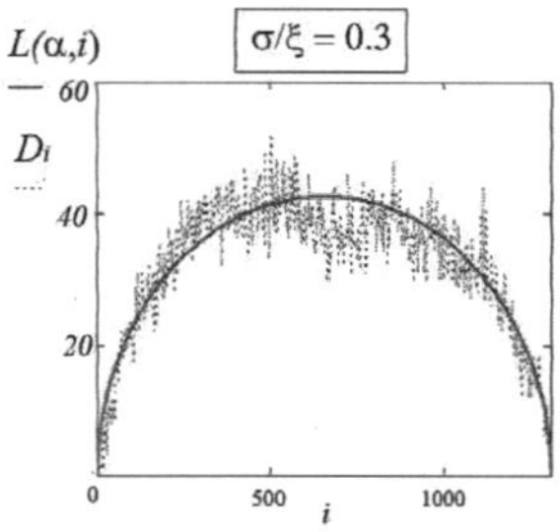


(a)

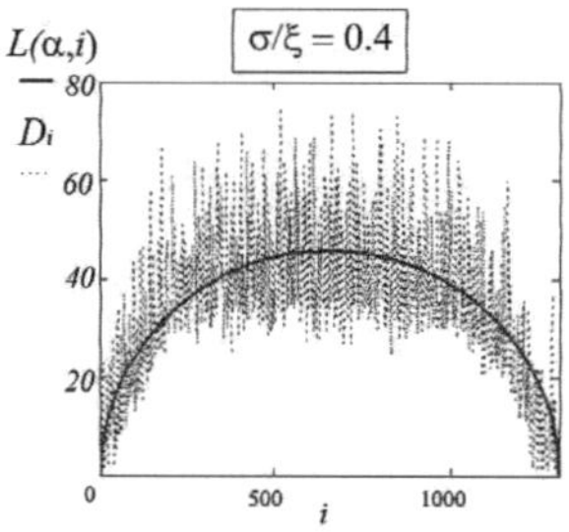


(b)

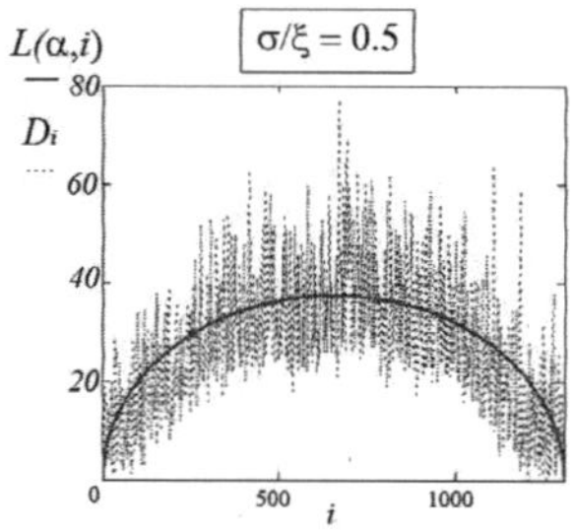


(c)

**Figure 3**. Brightness distribution in the central layer of the air gap (10 μm). $L(\alpha,i)$ – ideal brightness distribution, $D_i$ – experimental brightness values. $\sigma$ – standard deviation, $\xi$ – average brightness value. a) $d = 1$ mm, b) $d = 2$ mm, c) $d = 3$ mm.

The minimum value of $\sigma/\xi = 0.3$ is observed in 1-mm air gaps. With the height increasing this ratio also increases. Thus, it can be concluded that the degree of NDBD homogeneity got worse with increasing the air gap.

The ratio $E/p$ in the air gap, and the growth rate of electric field strength in the air gap $dE/dt$ appeared to play the major role in the NDBD development, as the gas composition and parameters of the discharge gap and the external circuit were the same during all experiments.

The growth rate of electric field strength in the discharge gap $dE/dt$ was estimated to study the causes of various discharge behavior depending on the height. For the calculations we used the value of the voltage at the electrodes at the ignition moment, i.e. at the moment of the discharge current appearance.

NDBD was ignited at different voltage across the electrodes depending on the height of the air gap. The voltage amplitude across the electrodes at the ignition moment was 10.8 kV for 1-mm gap, 14.5 kV – for 2 mm gap and 19 kV – for 3-mm gap.

The equivalent circuit of the discharge gap in the absence of the discharge is a voltage divider: barrier capacity and "air gap" capacity, connected in series [10–12]. Therefore, the ignition voltage at the gap can be calculated taking into account factor $k$, defined as:

$$k = \frac{C_b}{C_a + C_b}, \qquad (2)$$

where $C_b$ – barrier capacity and $C_a$ – "air gap" capacity. In our experiments $C_b$ = 15 пФ, $C_a$ decreased form 6 pF to 2 pF depending on the height of the discharge gap $d$.

Thus, the ration $E/p$ can be estimated by the following expression:

$$\frac{E}{p} = \frac{kV_{in}}{pd}. \qquad (3)$$

It is easy to estimate the growth rate of electric field strength in the air gap with known $k$ and growth rate of the applied to the electrodes voltage $dV/dt \approx 250$ V/ns, which is determined by the peculiarities of the HVPG operation [8]. It can be calculated for a flat gap by the formula:

$$\frac{dE}{dt} = \frac{dV}{dt} \cdot \frac{k}{d}. \qquad (4)$$

Table 1 shows the values of $E/p$ and the growth rate of electric field strength in the air gap for three values of the gap heights 1, 2 and 3 mm, the NDBD was realized at.

**Table 1.** The values of $E/p$ and the growth rate of electric field strength in the discharge gap

| ***d*** | ***E/p*** | **d*E*/d*t*** |
|---|---|---|
| 1 mm | 96 V/(Torr·cm) | 1.7 kV/(cm·ns) |
| 2 mm | 80 V/(Torr·cm) | 1.0 kV/(cm·ns) |
| 3 mm | 71 V/(Torr·cm) | 0.7 kV/(cm·ns) |

As it can be seen from the data, presented in table 1, the greatest $E/p$ and growth rate of the electric field strength in the air gap have been obtained in 1-mm gaps.

## 4. Summary

According to the obtained experimental data it was shown that the NDBD in millimeter air gaps under atmospheric pressure and natural humidity developed in various ways depending on the height of the air gap, when applying to the electrodes the voltage pulses with rise time 40 ns. It was found out that the higher degree of homogeneity of the NDBD corresponded to the rather synchronous development of the discharge currents in the gap with delay between them no more than 1 ns. It was obtained from the experimental data that $E/p$ and the growth rate of electric field in the air gap $dE/dt$ in this regime had the highest value for the investigated range of air gaps. It was shown that at $E/p \geq 80$ V/(Torr·cm) and $dE/dt \geq 1$ V/(Torr·cm) the volume diffuse NDBD was realized in 1-mm and 2-mm air gaps, while at a lower value $dE/dt$ and $E/p < 80$ V/(Torr·cm) the existence of the volume diffuse NDBD in millimeter air gaps is difficult, as it has been confirmed experimentally for $d = 3$ mm, when $dE/dt \approx 0.7$ kV/(cm·ns).

**Acknowledgments**

This work was partially supported by the Russian Foundation for Basic Research, grant No. 16-08-01037A.

**References**

[1] Shao T, Zhang D, Yu Y, Zhang C, Wang J, Yan P, Zhou Y 2010 *Plasma Science, IEEE Transactions on* **38** 1651

[2] Ayan H, Fridman G, Gutsol A F, Vasilets V N, Fridman A, Friedman G 2008 *Plasma Science, IEEE Transactions on* **36** 504

[3] Khomich V Y, Malashin M V, Moshkunov S I, Shershunova E A, Yamschikov V A 2014 *Plasma Science, IEEE Transactions on 42* 3314

[4] Shershunova E, Malashin M, Moshkunov S, Khomich V Generation of Homogeneous Dielectric Barrier Discharge in Atmospheric Air without Preionization 2013 *19th International Conference on Plasma Processes: Abstract Book IVC-19. T11. Plasma Science and Technology. PST–P2-02* (Paris, France, September 9–13 2013) 1242

[5] Shao T, Zhang C, Yu Y, Fang Z, Yan P 2012 *Europhysics Letters* **97** 55005

[6] Mangolini L, Orlov K, Kortshagen U, Heberlein J, Kogelschatz U 2002 **80** 1722

[7] Malashin M V, Moshkunov S I, Khomich V Y 2016 *Plasma Physics Reports* **42** 163

[8] Malashin M V, Moshkunov S I, Khomich V Y, Shershunova E A 2016 *Instruments and Experimental Techniques* **59** 226

[9] Malashin M V, Moshkunov S I, Khomich V Y, Shershunova E A 2015 *Technical Physics Letters* **41** 436

[10] Laroussi M, Lu X, Kolobov V, Arslanbekov R 2004 *Journal of Applied Physics* **96** 3028

[11] Shershunova E, Malashin M, Moshkunov S, Khomich V 2015 *Acta Polytechnica* **55** 59

[12] Pipa A V, Koskulics J, Brandenburg R, Hoder T 2012 *Review of Scientific Instruments* **83** 115112